Atlantis Studies in Scientific Computing in Electromagnetics

Volume 1

Series editor

Wil Schilders, Technische Universiteit Eindhoven, Eindhoven, The Netherlands

For further volumes:
http://www.atlantis-press.com/series/13301

This series contains volumes on scientific computing for a wide range of electromagnetics problems. The electronics industry, in a very broad sense, is at the forefront of innovation, and our daily life is very much dependent on achievements in this area. These are mainly enabled by rapid developments in sophisticated virtual design environments, numerical methods being at the core of these. Volumes in the series provide details on the modeling, analysis and simulation of problems, as well as on the design process of robust solution methods. Applications range from simple benchmark problems to industrial problems at the forefront of current research and development.

For more information on this series and our other book series, please visit our website: www.atlantis-press.com

Atlantis Press
8 square des Bouleaux
75019 Paris, France

Reijer Idema · Domenico J. P. Lahaye

Computational Methods in Power System Analysis

ATLANTIS PRESS

Reijer Idema
Domenico J. P. Lahaye
Numerical Analysis
Delft University of Technology
Delft
The Netherlands

ISSN 2352-0590 ISSN 2352-0604 (electronic)
ISBN 978-94-6239-063-8 ISBN 978-94-6239-064-5 (eBook)
DOI 10.2991/978-94-6239-064-5

Library of Congress Control Number: 2013957992

Printed on acid-free paper

Preface

There are many excellent books on power systems that treat power system analysis, and its most important computational problem: the power flow problem. Some of these books also discuss the traditional computational methods for solving the power flow problem, i.e., Newton power flow and Fast Decoupled Load Flow. However, information on newer solution methods is hard to find outside research papers.

This book aims to fill that gap, by offering a self-contained volume that treats both traditional and newer methods. It is meant both for researchers who want to get into the subject of power flow and related problems, and for software developers that work on power system analysis tools.

Part I of the book treats the mathematics and computational methods needed to understand modern power flow methods. Depending on the knowledge and interest of the reader, it can be read in its entirety or used as a reference when reading Part II. Part II treats the application of these computational methods to the power flow problem and related power system analysis problems, and should be considered the meat of this publication.

This book is based on research conducted by the authors at the Delft University of Technology, in collaboration between the Numerical Analysis group of the Delft Institute of Applied Mathematics and the Electrical Power Systems group, both in the faculty Electrical Engineering, Mathematics and Computer Science.

The authors would like to acknowledge Kees Vuik, the Numerical Analysis chair, and Lou van der Sluis, the Electrical Power Systems chair, for the fruitful collaboration, as well as all colleagues of both groups that had a part in our research. Special thanks are extended to Robert van Amerongen, who was vital in bridging the gap between applied mathematics and electrical engineering.

Further thanks go to Barry Smith of the Argonne National Laboratory for his help with the PETSc package, and ENTSO-E for providing the UCTE study model.

Delft, October 2013

Reijer Idema
Domenico J. P. Lahaye

Preface

Contents

Chapter 1
Introduction

Electricity is a vital part of modern society. We plug our electronic devices into wall sockets and expect them to get power. Power generation is a subject that is in the news regularly. The issue of the depletion of natural resources and the risks of nuclear power plants are often discussed, and developments in wind and solar power generation, as well as other renewables, are hot topics. Much less discussed is the transmission and distribution of electrical power, an incredibly complex task that needs to be executed reliably and securely, and highly efficiently. To achieve this, both operation and planning require many complex computational simulations of the power system network.

Traditionally, power generation is centralised in large plants that are connected directly to the transmission system. The high voltage transmission system transports the generated power to the lower voltage local distribution systems. In recent years, decentralised power generation has been emerging, for example in the form of solar panels on the roofs of residential houses, or small wind farms that are connected to the distribution network. It is expected that the future will bring a much more decentralised power system. This leads to many new computational challenges in power system operation and planning.

Meanwhile, national power systems are being interconnected more and more, and with it the associated energy markets. The resulting continent-wide power systems lead to much larger power system simulations.

The base computational problem in steady-state power system simulations is the power flow (or load flow) problem. The power flow problem is a nonlinear system of equations that relates the bus voltages to the power generation and consumption. For given generation and consumption, the power flow problem can be solved to reveal the associated bus voltages. The solution can be used to assess whether the power system will function properly. Power flow studies are the main ingredient of many computations in power system analysis.

Contingency analysis simulates equipment outages in the power system, and solves the associated power flow problems to assess the impact on the power system. Contingency analysis is vital to identify possible problems, and solve them before

R. Idema and D. J. P. Lahaye, *Computational Methods in Power System Analysis*,
Atlantis Studies in Scientific Computing in Electromagnetics,
DOI: 10.2991/978-94-6239-064-5_1,

they have a chance to occur. Many countries require their power system to operate in such a way that no single equipment outage causes interruption of service.

Monte Carlo simulations, with power flow calculations for many varying generation and consumption inputs, can be used to analyse the stochastic behaviour of a power system. This type of simulation is becoming especially important due to the uncontrollable nature of wind and solar power.

Operation and planning of power systems further lead to many kinds of optimisation problems. What power plants should be generating how much power at any given time? Where to best build a new power plant? Which buses to connect with a new line or cable? All these questions require the solution of some optimisation problem, where the set of feasible solutions is determined by power flow problems, or even contingency analysis and Monte Carlo simulations.

Traditionally, the power flow problem is solved using Newton power flow or the Fast Decoupled Load Flow (FDLF) method. Newton power flow has the quadratic convergence behaviour of the Newton-Raphson method, but needs a lot of computational work per iteration, especially for large power flow problems. FDLF needs relatively little computational work per iteration, but the convergence is only linear. In practice, Newton power flow is generally preferred because it is more robust, i.e., for some power flow problems FDLF fails to converge, while Newton power flow can still solve the problem. However, neither method is viable for very large power flow problems. Therefore, the development of fast and scalable power flow solvers is very important for the continuous operation of future power systems.

In this book, Newton-Krylov power flow solvers are treated that are as fast as traditional solvers for small power flow problems, and many times faster for large problems. Further, contingency analysis is used to demonstrate how these solvers can be used to speed up the computation of many slightly varying power flow problems, as found not only in contingency analysis, but also in Monte Carlo simulations and some optimisation problems.

In Part I the relevant computational methods are treated. The theory behind solvers for linear and nonlinear systems of equations is treated to provide a solid understanding of Newton-Krylov methods, and convergence theory is discussed, as it is needed to be able to make the right choices for the Krylov method, preconditioning, and forcing terms, and to correctly interpret the convergence behaviour of numerical experiments.

In Part II power system analysis is treated. The relevant power system theory is described, traditional solvers are explained in detail, and Newton-Krylov power flow solvers are discussed and tested, using many combinations of choices for the Krylov method, preconditioning, and forcing terms.

It is explained that Newton power flow and FDLF can be seen as elementary Newton-Krylov methods, indicating that the developed Newton-Krylov power flow solvers are a direct theoretical improvement on these traditional solvers. It is shown, both theoretically and experimentally, that well-designed Newton-Krylov power flow solvers have no drawbacks in terms of speed and robustness, while scaling much better in the problem size, and offering even more computational advantage when solving many slightly varying power flow problems.

Part I
Computational Methods

Chapter 2
Fundamental Mathematics

This chapter gives a short introduction to fundamental mathematical concepts that are used in the computational methods treated in this book. These concepts are complex numbers, vectors, matrices, and graphs. Vectors and matrices belong to the field of linear algebra. For more information on linear algebra, see for example [1], which includes an appendix on complex numbers. For more information on spectral graph theory, see for example [2].

2.1 Complex Numbers

A complex number $\alpha \in \mathbb{C}$, is a number

$$\alpha = \mu + \iota\nu, \tag{2.1}$$

with $\mu, \nu \in \mathbb{R}$, and ι the imaginary unit[1] defined by $\iota^2 = -1$. The quantity $\operatorname{Re}\alpha = \mu$ is called the real part of α, whereas $\operatorname{Im}\alpha = \nu$ is called the imaginary part of the complex number. Note that any real number can be interpreted as a complex number with the imaginary part equal to 0.

Negation, addition, and multiplication are defined as

$$-(\mu + \iota\nu) = -\mu - \iota\nu, \tag{2.2}$$

$$\mu_1 + \iota\nu_1 + \mu_2 + \iota\nu_2 = (\mu_1 + \mu_2) + \iota(\nu_1 + \nu_2), \tag{2.3}$$

$$(\mu_1 + \iota\nu_1)(\mu_2 + \iota\nu_2) = (\mu_1\mu_2 - \nu_1\nu_2) + \iota(\mu_1\nu_2 + \mu_2\nu_1). \tag{2.4}$$

---

[1] The imaginary unit is usually denoted by i in mathematics, and by j in electrical engineering because i is reserved for the current. In this book, the imaginary unit is sometimes part of a matrix or vector equation where i and j are used as indices. To avoid ambiguity, the imaginary unit is therefore denoted by ι (iota).

R. Idema and D. J. P. Lahaye, *Computational Methods in Power System Analysis*, Atlantis Studies in Scientific Computing in Electromagnetics,
DOI: 10.2991/978-94-6239-064-5_2,

The complex conjugate is an operation that negates the imaginary part:

$$\overline{\mu + \iota\nu} = \mu - \iota\nu. \tag{2.5}$$

Complex numbers are often interpreted as points in complex plane, i.e., 2-dimensional space with a real and imaginary axis. The real and imaginary part are then the Cartesian coordinates of the complex point. That same point in the complex plane can also be described by an angle and a length. The angle of a complex number is called the argument, while the length is called the modulus:

$$\arg(\mu + \iota\nu) = \tan^{-1}\frac{\nu}{\mu}, \tag{2.6}$$

$$|\mu + \iota\nu| = \sqrt{\mu^2 + \nu^2}. \tag{2.7}$$

Using these definitions, any complex number $\alpha \in \mathbb{C}$ can be written as

$$\alpha = |\alpha|\, e^{\iota\varphi}, \tag{2.8}$$

where $\varphi = \arg\alpha$, and the complex exponential function is defined by

$$e^{\mu+\iota\nu} = e^{\mu}\left(\cos\nu + \iota\sin\nu\right). \tag{2.9}$$

2.2 Vectors

A vector $\mathbf{v} \in K^n$ is an element of the n-dimensional space of either real numbers ($K = \mathbb{R}$) or complex numbers ($K = \mathbb{C}$), generally denoted as

$$\mathbf{v} = \begin{bmatrix} v_1 \\ \vdots \\ v_n \end{bmatrix}, \tag{2.10}$$

where $v_1, \ldots, v_n \in K$.

Scalar multiplication and vector addition are basic operations that are performed elementwise. That is, for $\alpha \in K$ and $\mathbf{v}, \mathbf{w} \in K^n$,

$$\alpha\mathbf{v} = \begin{bmatrix} \alpha v_1 \\ \vdots \\ \alpha v_n \end{bmatrix}, \quad \mathbf{v} + \mathbf{w} = \begin{bmatrix} v_1 + w_1 \\ \vdots \\ v_n + w_n \end{bmatrix}. \tag{2.11}$$

The combined operation of the form $\mathbf{v} := \alpha\mathbf{v} + \beta\mathbf{w}$ is known as a vector update. Vector updates are of $O(n)$ complexity, and are naturally parallelisable.

A linear combination of the vectors $\mathbf{v}_1, \ldots, \mathbf{v}_m \in K^n$ is an expression

$$\alpha_1 \mathbf{v}_1 + \ldots + \alpha_m \mathbf{v}_m, \tag{2.12}$$

with $\alpha_1 \ldots \alpha_m \in K$. A set of m vectors $\mathbf{v}_1, \ldots, \mathbf{v}_m \in K^n$ is called linearly independent, if none of the vectors can be written as a linear combination of the other vectors.

The dot product operation is defined for real vectors $\mathbf{v}, \mathbf{w} \in \mathbb{R}^n$ as

$$\mathbf{v} \cdot \mathbf{w} = \sum_{i=1}^{n} v_i w_i. \tag{2.13}$$

The dot product is by far the most used type of inner product. In this book, whenever we speak of an inner product, we will be referring to the dot product unless stated otherwise. The operation is of $O(n)$ complexity, but not naturally parallelisable. The dot product can be extended to complex vectors $\mathbf{v}, \mathbf{w} \in \mathbb{C}$ as $\mathbf{v} \cdot \mathbf{w} = \sum_{i=1}^{n} \overline{v}_i w_i$.

A vector norm is a function $\|.\|$ that assigns a measure of length, or size, to all vectors, such that for all $\alpha \in K$ and $\mathbf{v}, \mathbf{w} \in K^n$

$$\|\mathbf{v}\| = 0 \Leftrightarrow \mathbf{v} = \mathbf{0}, \tag{2.14}$$

$$\|\alpha \mathbf{v}\| = |\alpha| \, \|\mathbf{v}\|, \tag{2.15}$$

$$\|\mathbf{v} + \mathbf{w}\| \leq \|\mathbf{v}\| + \|\mathbf{w}\|. \tag{2.16}$$

Note that these properties ensure that the norm of a vector is never negative. For real vectors $\mathbf{v} \in \mathbb{R}^n$ the Euclidean norm, or 2-norm, is defined as

$$\|\mathbf{v}\|_2 = \sqrt{\mathbf{v} \cdot \mathbf{v}} = \sqrt{\sum_{i=1}^{n} v_i^2}. \tag{2.17}$$

In Euclidean space of dimension n, the Euclidean norm is the distance from the origin to the point $\mathbf{v}$. Note the similarity between the Euclidean norm of a 2-dimensional vector and the modulus of a complex number. In this book we omit the subscripted 2 from the notation of Euclidean norms, and simply write $\|\mathbf{v}\|$.

2.3 Matrices

A matrix $A \in K^{m \times n}$ is a rectangular array of real numbers ($K = \mathbb{R}$) or complex numbers ($K = \mathbb{C}$), i.e.,

$$A = \begin{bmatrix} a_{11} & \dots & a_{1n} \\ \vdots & \ddots & \vdots \\ a_{m1} & \dots & a_{mn} \end{bmatrix}, \tag{2.18}$$

with $a_{ij} \in K$ for $i \in \{1, \dots, m\}$ and $j \in \{1, \dots, n\}$.

A matrix of dimension $n \times 1$ is a vector, sometimes referred to as a column vector to distinguish it from a matrix of dimension $1 \times n$, which is referred to as a row vector. Note that the columns of a matrix $A \in K^{m \times n}$ can be interpreted as n (column) vectors of dimension m, and the rows as m row vectors of dimension n.

A dense matrix is a matrix that contains mostly nonzero values; all n^2 values have to be stored in memory. If most values are zeros the matrix is called sparse. For a sparse matrix A, the number of nonzero values is denoted by $\mathrm{nnz}(A)$. With special data structures, only the $\mathrm{nnz}(A)$ nonzero values have to be stored in memory.

The transpose of a matrix $A \in K^{m \times n}$, is the matrix $A^T \in K^{n \times m}$ with

$$\left(A^T\right)_{ij} = (A)_{ji} \, . \tag{2.19}$$

A square matrix that is equal to its transpose is called a symmetric matrix.

Scalar multiplication and matrix addition are elementwise operations, as with vectors. Let $\alpha \in K$ be a scalar, and $A, B \in K^{m \times n}$ matrices with columns $\mathbf{a}_i, \mathbf{b}_i \in K^m$ respectively, then scalar multiplication and matrix addition are defined as

$$\alpha \mathbf{A} = \begin{bmatrix} \alpha \mathbf{a}_1 & \dots & \alpha \mathbf{a}_n \end{bmatrix}, \tag{2.20}$$

$$\mathbf{A} + \mathbf{B} = \begin{bmatrix} \mathbf{a}_1 + \mathbf{b}_1 & \dots & \mathbf{a}_n + \mathbf{b}_n \end{bmatrix}. \tag{2.21}$$

Matrix-vector multiplication is the product of a matrix $A \in K^{m \times n}$ and a vector $\mathbf{v} \in K^n$, defined by

$$\begin{bmatrix} a_{11} & \dots & a_{1n} \\ \vdots & \ddots & \vdots \\ a_{m1} & \dots & a_{mn} \end{bmatrix} \begin{bmatrix} v_1 \\ \vdots \\ v_n \end{bmatrix} = \begin{bmatrix} \sum_{i=1}^{n} a_{1i} v_i \\ \vdots \\ \sum_{i=1}^{n} a_{mi} v_i \end{bmatrix}. \tag{2.22}$$

Note that the result is a vector in K^m. An operation of the form $\mathbf{u} := A\mathbf{v}$ is often referred to as a matvec. A matvec with a dense matrix has $O(n^2)$ complexity, while with a sparse matrix the operation has $O(\mathrm{nnz}(A))$ complexity. Both dense and sparse versions are naturally parallelisable.

Multiplication of matrices $A \in K^{m \times p}$ and $B \in K^{p \times n}$ can be derived as an extension of matrix-vector multiplication by writing the columns of B as vectors $\mathbf{b}_i \in K^p$. This gives

$$\begin{bmatrix} a_{11} & \dots & a_{1n} \\ \vdots & \ddots & \vdots \\ a_{m1} & \dots & a_{mn} \end{bmatrix} \begin{bmatrix} \mathbf{b}_1 \dots \mathbf{b}_n \end{bmatrix} = \begin{bmatrix} A\mathbf{b}_1 \dots A\mathbf{b}_n \end{bmatrix}. \tag{2.23}$$

The product AB is a matrix of dimension $m \times n$.

The identity matrix I is the matrix with values $I_{ii} = 1$, and $I_{ij} = 0$, $i \neq j$. Or, in words, the identity matrix is a diagonal matrix with every diagonal element equal to 1. This matrix is such, that $IA = A$ and $AI = A$ for any matrix $A \in K^{m \times n}$ and identity matrices I of appropriate size.

Let $A \in K^{n \times n}$ be a square matrix. If there is a matrix $B \in K^{n \times n}$ such that $BA = I$, then B is called the inverse of A. If the inverse matrix does not exist, then A is called singular. If it does exist, then it is unique and denoted by A^{-1}. Calculating the inverse has $O(n^3)$ complexity, and is therefore very costly for large matrices.

The column rank of a matrix $A \in K^{m \times n}$ is the number of linearly independent column vectors in A. Similarly, the row rank is the number of linearly independent row vectors in A. For any given matrix, the row rank and column rank are equal, and can therefore simply be denoted as rank (A). A square matrix $A \in K^{n \times n}$ is invertible, or nonsingular, if and only if rank $(A) = n$.

A matrix norm is a function $\|.\|$ such that for all $\alpha \in K$ and $A, B \in K^{m \times n}$

$$\|A\| \geq 0, \tag{2.24}$$

$$\|\alpha A\| = |\alpha| \, \|A\|, \tag{2.25}$$

$$\|A + B\| \leq \|A\| + \|B\|. \tag{2.26}$$

Given a vector norm $\|.\|$, the corresponding induced matrix norm is defined for all matrices $A \in K^{m \times n}$ as

$$\|A\| = \max \left\{ \|A\mathbf{v}\| : \mathbf{v} \in K^n \text{ with } \|\mathbf{v}\| = 1 \right\}. \tag{2.27}$$

Every induced matrix norm is submultiplicative, meaning that

$$\|AB\| \leq \|A\| \|B\| \text{ for all } A \in K^{m \times p}, \; B \in K^{p \times n}. \tag{2.28}$$

2.4 Graphs

A graph is a collection of vertices, any pair of which may be connected by an edge. Vertices are also called nodes or points, and edges are also called lines. The graph is called directed if all edges have a direction, and undirected if they do not. Graphs are often used as the abstract representation of some sort of network. For example, a power system network can be modelled as an undirected graph, with buses as vertices and branches as edges.

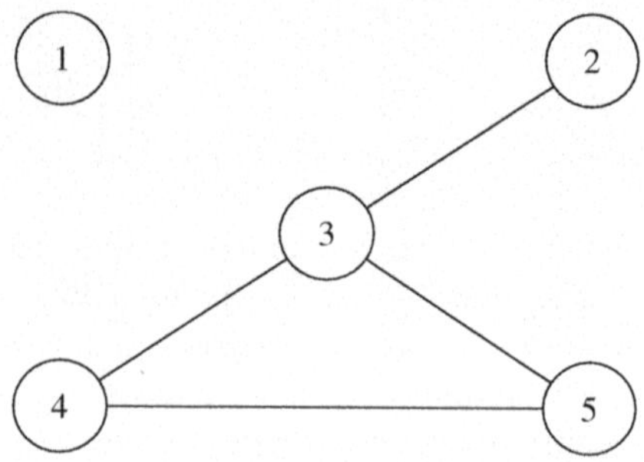

Fig. 2.1 A simple graph

Let $V = \{v_1, \ldots, v_N\}$ be a set of N vertices, and $E = \{e_1, \ldots, e_M\}$ a set of M edges, where each edge $e_k = \left(v_i, v_j\right)$ connects two vertices $v_i, v_j \in V$. The graph G of vertices V and edges E is denoted as $G = (V, E)$. Figure 2.1 shows a graph $G = (V, E)$ with vertices $V = \{1, 2, 3, 4, 5\}$ and edges $E = \{(2, 3), (3, 4), (3, 5), (4, 5)\}$.

The incidence matrix A of a graph $G = (V, E)$ is an $M \times N$ matrix in which each row i represents an edge $e_i = (p, q)$, and is defined as

$$a_{ij} = \begin{cases} -1 \text{ if } p = v_i, \\ 1 \text{ if } q = v_j, \\ 0 \text{ otherwise.} \end{cases} \tag{2.29}$$

In other words, row i has value -1 at index p and value 1 at index q. Note that this matrix is unique for a directed graph. For an undirected graph, some orientation has to be chosen. For example, the matrix

$$A = \begin{bmatrix} 0 & -1 & 1 & 0 & 0 \\ 0 & 0 & -1 & 1 & 0 \\ 0 & 0 & -1 & 0 & 1 \\ 0 & 0 & 0 & -1 & 1 \end{bmatrix} \tag{2.30}$$

is an incidence matrix of the graph in Fig. 2.1. Such a matrix is sometimes referred to as an oriented incidence matrix, to distinguish it from the unique unoriented incidence matrix, in which all occurrences of -1 are replaced with 1. Note that some authors define the incidence matrix as the transpose of the matrix A defined here.

References

1. Lay, D.C.: Linear Algebra And Its Applications, 4th edn. Pearson Education, Toronto (2011)
2. Chung, F.R.K.: Spectral Graph Theory. No. 92 in CBMS Regional Conference Series. Conference Board of the Mathematical Sciences, Washington (1997)

Chapter 3
Solving Linear Systems of Equations

A linear equation in n variables $x_1, \ldots, x_n \in \mathbb{R}$, is an equation of the form

$$a_1 x_1 + \cdots + a_n x_n = b, \tag{3.1}$$

with given constants $a_1, \ldots, a_n, b \in \mathbb{R}$. If there is at least one coefficient a_i not equal to 0, then the solution set is an $(n-1)$-dimensional affine hyperplane in $\mathbb{R}^n$. If all coefficients are equal to 0, then there is either no solution if $b \neq 0$, or the solution set is the entire space $\mathbb{R}^n$ if $b = 0$.

A linear system of equations is a collection of linear equations in the same variables that have to be satisfied simultaneously. Any linear system of m equations in n variables can be written as

$$A\mathbf{x} = \mathbf{b}, \tag{3.2}$$

where $A \in \mathbb{R}^{m \times n}$ is called the coefficient matrix, $\mathbf{b} \in \mathbb{R}^m$ the right-hand side vector, and $\mathbf{x} \in \mathbb{R}^n$ the vector of variables or unknowns.

If there exists at least one solution vector $\mathbf{x}$ that satisfies all linear equations at the same time, then the linear system is called consistent; otherwise, it is called inconsistent. If the right-hand side vector $\mathbf{b} = \mathbf{0}$, then the system of equations is always consistent, because the trivial solution $\mathbf{x} = \mathbf{0}$ satisfies all equations independent of the coefficient matrix.

We focus on systems of linear equations with a square coefficient matrix:

$$A\mathbf{x} = \mathbf{b}, \text{ with } A \in \mathbb{R}^{n \times n} \text{ and } \mathbf{b}, \mathbf{x} \in \mathbb{R}^n. \tag{3.3}$$

If all equations are linearly independent, i.e., if $\text{rank}(A) = n$, then the matrix A is invertible and the linear system (3.3) has a unique solution $\mathbf{x} = A^{-1}\mathbf{b}$. If not all equations are linearly independent, i.e., if $\text{rank}(A) < n$, then A is singular. In this case the system is either inconsistent, or the solution set is a subspace of dimension $n - \text{rank}(A)$. Note that whether there is exactly one solution or not can be deduced from the coefficient matrix alone, while both coefficient matrix and right-hand side vector are needed to distinguish between no solutions or infinitely many solutions.

R. Idema and D. J. P. Lahaye, *Computational Methods in Power System Analysis*, Atlantis Studies in Scientific Computing in Electromagnetics,
DOI: 10.2991/978-94-6239-064-5_3,

A solver for systems of linear equations can either be a direct method, or an iterative method. Direct methods calculate the solution to the problem in one pass. Iterative methods start with some initial vector, and update this vector in every iteration until it is close enough to the solution. Direct methods are very well-suited for smaller problems, and for problems with a dense coefficient matrix. For large sparse problems, iterative methods are generally much more efficient than direct solvers.

3.1 Direct Solvers

A direct solver may consist of a method to calculate the inverse coefficient matrix A^{-1}, after which the solution of the linear system (3.3) can simply be found by calculating the matvec $\mathbf{x} = A^{-1}\mathbf{b}$. In practice, it is generally more efficient to build a factorisation of the coefficient matrix into triangular matrices, which can then be used to easily derive the solution. For general matrices the factorisation of choice is the LU decomposition.

3.1.1 LU Decomposition

The LU decomposition consists of a lower triangular matrix L, and an upper triangular matrix U, such that

$$LU = A. \tag{3.4}$$

The factors are unique if the requirement is added that all diagonal elements of either L or U are ones.

Using the LU decomposition, the system of linear equations (3.3) can be written as

$$LU\mathbf{x} = \mathbf{b}, \tag{3.5}$$

and solved by consecutively solving the two linear systems

$$L\mathbf{y} = \mathbf{b}, \tag{3.6}$$

$$U\mathbf{x} = \mathbf{y}. \tag{3.7}$$

Because L and U are triangular, these systems are quickly solved using forward and backward substitution respectively.

The rows and columns of the coefficient matrix A can be permutated freely without changing the solution of the linear system (3.3), as long as the vectors $\mathbf{b}$ and $\mathbf{x}$ are permutated accordingly. Using such permutations during the factorisation process is called pivoting. Allowing only row permutations during factorisation is often referred to as partial pivoting.

Every invertible matrix A has an LU decomposition if partial pivoting is allowed. For some singular matrices an LU decomposition also exists, but for many there is no such factorisation possible. In general, direct solvers have problems with solving linear systems with singular coefficient matrices.

More information on the LU decomposition can be found in [1–3].

3.1.2 Solution Accuracy

Direct solvers are often said to calculate the exact solution, unlike iterative solvers, which calculate approximate solutions. Indeed, the algorithms of direct solvers lead to an exact solution in exact arithmetic. However, although the algorithms may be exact, the computers that execute them are not. Finite precision arithmetic may still introduce errors in the solution calculated by a direct solver.

During the factorisation process, rounding errors may lead to substantial inaccuracies in the factors. Errors in the factors, in turn, lead to errors in the solution vector calculated by forward and backward substitution. Stability of the factorisation can be improved by using a good pivoting strategy during the process. The accuracy of the factors L and U can also be improved afterwards by simple iterative refinement techniques [2].

3.1.3 Algorithmic Complexity

Forward and backward substitution operations have complexity $O(\text{nnz}(A))$. For dense coefficient matrices, the complexity of the LU decomposition is $O(n^3)$. For sparse matrix systems, special sparse methods improve on this by exploiting the sparsity structure of the coefficient matrix. However, in general these methods still do not scale as well in the system size as iterative solvers can. Therefore, good iterative solvers will always be more efficient than direct solvers for very large sparse coefficient matrices.

To solve multiple systems of linear equations with the same coefficient matrix but different right-hand side vectors, it suffices to calculate the LU decomposition once at the start. Using this factorisation, the linear problem can be solved for each unique right-hand side by forward and backward substitution. Since the factorisation is far more time consuming than the substitution operations, this saves a lot of computational time compared to solving each linear system individually.

3.1.4 Fill-in and Matrix Ordering

In the LU decomposition of a sparse coefficient matrix A, there will be a certain amount of fill-in. Fill-in is the number of nonzero elements in L and U, of which the corresponding element in A is zero. Fill-in not only increases the amount of memory needed to store the factors, but also increases the complexity of the LU decomposition, as well as the forward and backward substitution operations.

The ordering of rows and columns—controlled by pivoting—can have a strong influence on the amount of fill-in. Finding the ordering that minimises fill-in has been proven to be NP-hard [4]. However, many methods have been developed that quickly find a good reordering, see for example [1, 5].

3.1.5 Incomplete LU decomposition

An incomplete LU decomposition [6, 7], or ILU decomposition, is a factorisation of A into a lower triangular matrix L, and an upper triangular matrix U, such that

$$LU \approx A. \tag{3.8}$$

The aim is to reduce computational cost by reducing the fill-in compared to the complete LU factors.

One method simply calculates the LU decomposition, and then drops all entries that are below a certain tolerance value. Obviously, this method does not reduce the complexity of the decomposition operation. However, the fill-in reduction saves memory, and reduces the computational cost of forward and backward substitution operations.

The ILU(k) method determines which entries in the factors L and U are allowed to be nonzero, based on the number of levels of fill $k \in \mathbb{N}$. ILU(0) is an incomplete LU decomposition such that $L + U$ has the same nonzero pattern as the original matrix A. For sparse matrices, this method is often much faster than the complete LU decomposition.

With an ILU(k) factorisation, the row and column ordering of A may still influence the number of nonzeros in the factors, although much less drastically than with the LU decomposition. Further, it has been observed that the ordering also influences the quality of the approximation of the original matrix. A reordering that reduces the fill-in often also reduces the approximation error of the ILU(k) factorisation.

It is clear that ILU factorisations are not suitable to be used in a direct solver, unless the approximation is very close to the original. In general, there is no point in using an ILU decomposition over the LU decomposition unless only a rough approximation of A is needed. ILU factorisations are often used a preconditioners for iterative linear solvers, see Sect. 3.2.4.

3.2 Iterative Solvers

Iterative solvers start with an initial iterate $\mathbf{x}_0$, and calculate a new iterate in each step, or iteration, thus producing a sequence of iterates $\{\mathbf{x}_0, \mathbf{x}_1, \mathbf{x}_2, \ldots\}$. The aim is that at some iteration i, the iterate $\mathbf{x}_i$ will be close enough to the solution to be used as approximation of the solution. When $\mathbf{x}_i$ is close enough to the solution, the method is said to have converged. Since the true solution is not known, $\mathbf{x}_i$ cannot simply be compared with that solution to decide if the method has converged; a different measure of the error in the iterate $\mathbf{x}_i$ is needed.

The residual vector in iteration i is defined by

$$\mathbf{r}_i = \mathbf{b} - A\mathbf{x}_i. \tag{3.9}$$

Let $\mathbf{e}_i$ denote the difference between $\mathbf{x}_i$ and the true solution. Then the norm of the residual is

$$\|\mathbf{r}_i\| = \|\mathbf{b} - A\mathbf{x}_i\| = \|A\mathbf{e}_i\| = \|\mathbf{e}_i\|_{A^T A}. \tag{3.10}$$

This norm is a measure for the error in $\mathbf{x}_i$, and referred to as the residual error. The relative residual norm $\frac{\|\mathbf{r}_i\|}{\|\mathbf{b}\|}$ can be used as a measure of the relative error in the iterate $\mathbf{x}_i$.

3.2.1 Krylov Subspace Methods

The Krylov subspace of dimension i, belonging to A and $\mathbf{r}_0$, is defined as

$$\mathcal{K}_i(A, \mathbf{r}_0) = \text{span}\{\mathbf{r}_0, A\mathbf{r}_0, \ldots, A^{i-1}\mathbf{r}_0, \}. \tag{3.11}$$

Krylov subspace methods are iterative linear solvers that generate iterates

$$\mathbf{x}_i \in \mathbf{x}_0 + \mathcal{K}_i(A, \mathbf{r}_0). \tag{3.12}$$

The simplest Krylov method consists of the Richardson iterations,

$$\mathbf{x}_{i+1} = \mathbf{x}_i + \mathbf{r}_i. \tag{3.13}$$

Basic iterative methods like Jacobi, Gauss-Seidel, and Successive Over-Relaxation (SOR) iterations, can all be seen as preconditioned versions of the Richardson iterations. Preconditioning is treated in Sect. 3.2.4. More information on basic iterative methods can be found in [2, 8, 9].

Krylov subspace methods generally have no problem finding a solution for a consistent linear system with a singular coefficient matrix A. Indeed, the dimension of the Krylov subspace needed to describe the full column space of A is equal to $\text{rank}(A)$, and is therefore lower for singular matrices than for invertible matrices.

Popular iterative linear solvers for general square coefficient matrices include GMRES [10], Bi-CGSTAB [11, 12], and IDR(s) [13]. These methods are more complex than the basic iterative methods, but generally converge a lot faster to a solution. All these iterative linear solvers can also be characterised as Krylov subspace methods. For an extensive treatment of Krylov subspace methods see [8].

3.2.2 Optimality and Short Recurrences

Two important properties of Krylov methods are the optimality property, and short recurrences. The first is about minimising the number of iterations needed to find a good approximation of the solution, while the second is about limiting the amount of computational work per iteration.

A Krylov method is said to have the optimality property, if in each iteration the computed iterate is the best possible approximation of the solution within current the Krylov subspace, i.e., if the residual norm $\|\mathbf{r}_i\|$ is minimised within the Krylov subspace. An iterative solver with the optimality property, is also called a minimal residual method.

An iterative process is said to have short recurrences if in each iteration only data from a small fixed number of previous iterations is used. If the needed amount of data and work keeps growing with the number of iterations, the algorithm is said to have long recurrences.

It has been proven that a Kylov method for general coefficient matrices can not have both the optimality property and short recurrences [14, 15]. As a result, the Generalised Minimal Residual (GMRES) method necessarily has long recurrences. Using restarts or truncation, GMRES can be made into a short recurrence method without optimality. Bi-CGSTAB and IDR(s) have short recurrences, but do not meet the optimality property.

3.2.3 Algorithmic Complexity

The matrix and vector operations that are used in Krylov subspace methods are generally restricted to matvecs, vector updates, and inner products. Of these operations, matvecs have the highest complexity with $O(\text{nnz}(A))$. Therefore, the complexity of Krylov methods is $O(\text{nnz}(A))$, provided convergence is reached in a limited number of iterations.

The computational work for a Krylov method is often measured in the number of matvecs, vector updates, and inner products used to increase the dimension of the Krylov subspace by one and find the new iterate within the expanded Krylov subspace. For short recurrence methods these numbers are fixed, while for long recurrences the computational work per iteration grows with the iteration count.

3.2.4 Preconditioning

No Krylov subspace method can produce iterates that are better than the best approximation of the solution within the progressive Krylov subspaces, which are the iterates attained by minimal residual methods. In other words, the convergence of a Krylov subspace method is limited by the Krylov subspace. Preconditioning uses a preconditioner matrix M to change the Krylov subspace, in order to improve convergence of the iterative solver.

Left Preconditioning

The system of linear equations (3.3) with left preconditioning becomes

$$M^{-1}A\mathbf{x} = M^{-1}\mathbf{b}. \tag{3.14}$$

The preconditioned residual for this linear system of equations is

$$\mathbf{r}_i = M^{-1}\left(\mathbf{b} - A\mathbf{x}_i\right), \tag{3.15}$$

and the new Krylov subspace is

$$\mathcal{K}_i(M^{-1}A, M^{-1}\mathbf{r}_0). \tag{3.16}$$

Right Preconditioning

The system of linear equations (3.3) with right preconditioning becomes

$$AM^{-1}\mathbf{y} = \mathbf{b}, \text{ and } \mathbf{x} = M^{-1}\mathbf{y}. \tag{3.17}$$

The preconditioned residual is the same as the unpreconditioned residual:

$$\mathbf{r}_i = \mathbf{b} - A\mathbf{x}_i. \tag{3.18}$$

The Krylov subspace for this linear system of equations is

$$\mathcal{K}_i(AM^{-1}, \mathbf{r}_0). \tag{3.19}$$

However, this Krylov subspace is used to generate iterates $\mathbf{y}_i$, which are not solution iterates like $\mathbf{x}_i$. Solution iterates $\mathbf{x}_i$ can be produced by multiplying $\mathbf{y}_i$ by M^{-1}. This leads to vectors $\mathbf{x}_i$ that are in the same Krylov subspace as with left preconditioning.

Split Preconditioning

Split preconditioning assumes some factorisation $M = M_L M_R$ of the preconditioner. The system of linear equations (3.3) then becomes

$$M_L^{-1}AM_R^{-1}\mathbf{y} = M_L^{-1}\mathbf{b}, \text{ and } \mathbf{x} = M_R^{-1}\mathbf{y}. \tag{3.20}$$

The preconditioned residual for this linear system of equations is

$$\mathbf{r}_i = M_L^{-1}\left(\mathbf{b} - A\mathbf{x}_i\right). \tag{3.21}$$

The Krylov subspace for the iterates $\mathbf{y}_i$ now is

$$\mathcal{K}_i(M_L^{-1}AM_R^{-1}, M_L^{-1}\mathbf{r}_0). \tag{3.22}$$

Transforming to solution iterates $\mathbf{x}_i = M_R^{-1}\mathbf{y}_i$, again leads to iterates in the same Krylov subspace as with left and right preconditioning.

Choosing the Preconditioner

Note that the explanation below assumes left preconditioning, but can easily be extended to right and split preconditioning.

To improve convergence, the preconditioner M needs to resemble the coefficient matrix A such that the preconditioned coefficient matrix $M^{-1}A$ resembles the identity matrix. At the same time, there should be a computationally cheap method available to evaluate $M^{-1}\mathbf{v}$ for any vector $\mathbf{v}$, because such an evaluation is needed in every preconditioned matvec in the Krylov subspace method.

A much used method is to create an LU decomposition of some matrix M that resembles A. In particular, an ILU decomposition of A can be used as preconditioner. With such a preconditioner it is important to control the fill-in of the factors, so that the overall complexity of the method does not increase much.

Another method of preconditioning, is to use an iterative linear solver to calculate a rough approximation of $\tilde{A}^{-1}\mathbf{v}$, and use this approximation instead of the explicit solution of $M^{-1}\mathbf{v}$. Here $\tilde{A}$ can be either the coefficient matrix A itself, or some convenient approximation of A. A stationary iterative linear solver can be used to precondition any Krylov subspace method, but nonstationary solvers require special flexible methods such as FGMRES [16].

3.2.5 *Starting and Stopping*

To start an iterative solver, an initial iterate $\mathbf{x}_0$ is needed. If some approximation of the solution of the linear system of equations is known, using it as initial iterate usually leads to fast convergence. If no such approximation is known, then usually the zero vector is chosen:

$$\mathbf{x}_0 = \mathbf{0}. \tag{3.23}$$

Another common choice is to use a random vector as initial iterate.

To stop the iteration process, some criterion is needed that indicates when to stop. By far the most common choice is to test if the relative residual error has become small enough, i.e., if for some choice of $\delta < 1$

$$\frac{\|\mathbf{r}_i\|}{\|\mathbf{b}\|} < \delta. \tag{3.24}$$

If left or split preconditioning is used, it is important to think about whether the true residual or the preconditioned residual should be used in the stopping criterion.

References

1. Duff, I.S., Erisman, A.M., Reid, J.K.: Direct Methods for Sparse Matrices. Oxford University Press, New York (1986)
2. Golub, G.H., van Loan, C.F.: Matrix Computations, 3rd edn. The Johns Hopkins University Press, Baltimore (1996)
3. Horn, R.A., Johnson, C.R.: Matrix Analysis, 3rd edn. Cambridge University Press, Cambridge (1990)
4. Yannakakis, M.: Computing the minimum fill-in is NP-complete. SIAM J. Algebraic Discrete Meth. **2**(1), 77–79 (1981)
5. Davis, T.A.: Direct Methods for Sparse Linear Systems. SIAM, Philadelphia (2006)
6. Meijerink, J.A., van der Vorst, H.A.: An iterative solution method for linear systems of which the coefficient matrix is a symmetric m-matrix. Math. Comput. **31**(137), 148–162 (1977)
7. Meijerink, J.A., van der Vorst, H.A.: Guidelines for the usage of incomplete decompositions in solving sets of linear equations as they occur in practical problems. J. Comput. Phys. **44**(1), 134–155 (1981)
8. Saad, Y.: Iterative methods for sparse linear systems, 2nd edn. SIAM, Philadelphia (2003)
9. Varga, R.S.: Matrix Iterative Analysis, 2nd edn. Springer, New York (2000)
10. Saad, Y., Schultz, M.H.: GMRES: a generalized minimal residual algorithm for solving non-symmetric linear systems. SIAM J. Sci. Stat. Comput. **7**, 856–869 (1986)
11. van der Vorst, H.A.: Bi-CGSTAB: a fast and smoothly converging variant of Bi-CG for solution of nonsymmetric linear systems. SIAM J. Sci. Stat. Comput. **13**, 631–644 (1992)
12. Sleijpen, G.L.G., van der Vorst, H.A., Fokkema, D.R.: BiCGstab(ℓ) and other hybrid Bi-CG methods. Numer. Algorithms **7**, 75–109 (1994)
13. Sonneveld, P., van Gijzen, M.B.: IDR(s): a family of simple and fast algorithms for solving large nonsymmetric systems of linear equations. SIAM J. Sci. Comput. **31**(2), 1035–1062 (2008)
14. Faber, V., Manteuffel, T.: Necessary and sufficient conditions for the existence of a conjugate gradient method. SIAM J. Numer. Anal. **21**, 352–362 (1984)
15. Voevodin, V.V.: The problem of non-self-adjoint generalization of the conjugate gradient method is closed. U.S.S.R. Comput. Math. Math. Phys. **22**, 143–144 (1983)
16. Saad, Y.: A flexible inner-outer preconditioned GMRES algorithm. SIAM J. Sci. Comput. **14**(2), 461–469 (1993)

Chapter 4
Solving Nonlinear Systems of Equations

A nonlinear equation in n variables $x_1, \ldots, x_n \in \mathbb{R}$, is an equation

$$f(x_1, \ldots, x_n) = 0, \tag{4.1}$$

that is not a linear equation.

A nonlinear system of equations is a collection of equations of which at least one equation is nonlinear. Any nonlinear system of m equations in n variables can be written as

$$\mathbf{F}(\mathbf{x}) = \mathbf{0}, \tag{4.2}$$

where $\mathbf{x} \in \mathbb{R}^n$ is the vector of variables or unknowns, and $\mathbf{F} : \mathbb{R}^n \to \mathbb{R}^m$ is a vector of m functions in $\mathbf{x}$, i.e.,

$$\mathbf{F}(\mathbf{x}) = \begin{bmatrix} F_1(\mathbf{x}) \\ \vdots \\ F_m(\mathbf{x}) \end{bmatrix}. \tag{4.3}$$

A solution of a nonlinear system of equations (4.2), is a vector $\mathbf{x}^* \in \mathbb{R}^n$ such that $F_k(\mathbf{x}^*) = 0$ for all $k \in \{1, \ldots, m\}$ at the same time. In this book, we restrict ourselves to nonlinear systems of equations with the same number of variables as there are equations, i.e., $m = n$.

It is not possible to solve a general nonlinear equation analytically, let alone a general nonlinear system of equations. However, there are iterative methods to find a solution for such systems. The Newton–Raphson algorithm is the standard method for solving nonlinear systems of equations. Most, if not all, other well-performing methods can be derived from the Newton–Raphson algorithm. In this chapter the Newton–Raphson method is treated, as well as some common variations.

R. Idema and D. J. P. Lahaye, *Computational Methods in Power System Analysis*,
Atlantis Studies in Scientific Computing in Electromagnetics,
DOI: 10.2991/978-94-6239-064-5_4,

4.1 Newton–Raphson Methods

The Newton–Raphson method is an iterative process used to solve nonlinear systems of equations

$$\mathbf{F}(\mathbf{x}) = \mathbf{0}, \tag{4.4}$$

where $\mathbf{F} : \mathbb{R}^n \rightarrow \mathbb{R}^n$ is continuously differentiable. In each iteration, the method solves a linearisation of the nonlinear problem around the current iterate, to find an update for that iterate. Algorithm 4.1 shows the basic Newton–Raphson process.

Algorithm 4.1 Newton–Raphson Method

1: $i := 0$
2: given initial iterate $\mathbf{x}_0$
3: **while** not converged **do**
4: solve $-J(\mathbf{x}_i)\mathbf{s}_i = \mathbf{F}(\mathbf{x}_i)$
5: update iterate $\mathbf{x}_{i+1} := \mathbf{x}_i + \mathbf{s}_i$
6: $i := i + 1$
7: **end while**

In Algorithm 4.1, the matrix J represents the Jacobian of $\mathbf{F}$, i.e.,

$$J = \begin{bmatrix} \frac{\partial F_1}{\partial x_1} & \cdots & \frac{\partial F_1}{\partial x_n} \\ \vdots & \ddots & \vdots \\ \frac{\partial F_n}{\partial x_1} & \cdots & \frac{\partial F_n}{\partial x_n} \end{bmatrix}. \tag{4.5}$$

The Jacobian system

$$-J(\mathbf{x}_i)\,\mathbf{s}_i = \mathbf{F}(\mathbf{x}_i) \tag{4.6}$$

can be solved using any linear solver. When a Krylov subspace method is used, we speak of a Newton–Krylov method.

The Newton process has local quadratic convergence. This means that if the iterate $\mathbf{x}_I$ is close enough to the solution, then there is a $c \geq 0$ such that for all $i \geq I$

$$\|\mathbf{x}_{i+1} - \mathbf{x}^*\| \leq c\|\mathbf{x}_i - \mathbf{x}^*\|^2. \tag{4.7}$$

The basic Newton method is not globally convergent, meaning that it does not always converge to a solution from every initial iterate $\mathbf{x}_0$. Line search and trust region methods can be used to augment the Newton method, to improve convergence if the initial iterate is far away from the solution, see Sect. 4.2.

As with iterative linear solvers, the distance of the current iterate to the solution is not known. The vector $\mathbf{F}(\mathbf{x}_i)$ can be seen as the nonlinear residual vector of iteration i. Convergence of the method is therefore mostly measured in the residual norm $\|\mathbf{F}(\mathbf{x}_i)\|$, or relative residual norm $\frac{\|\mathbf{F}(\mathbf{x}_i)\|}{\|\mathbf{F}(\mathbf{x}_0)\|}$.

4.1.1 Inexact Newton

Inexact Newton methods [1] are Newton–Raphson methods in which the Jacobian system (4.6) is not solved to full accuracy. Instead, in each Newton iteration the Jacobian system is solved such that

$$\frac{\|\mathbf{r}_i\|}{\|\mathbf{F}(\mathbf{x}_i)\|} \leq \eta_i, \tag{4.8}$$

where

$$\mathbf{r}_i = \mathbf{F}(\mathbf{x}_i) + J(\mathbf{x}_i)\,\mathbf{s}_i. \tag{4.9}$$

The values η_i are called the forcing terms.

The most common form of inexact Newton methods, is with an iterative linear solver to solve the Jacobian systems. The forcing terms then determine the accuracy to which the Jacobian system is solved in each Newton iteration. However, approximate Jacobian Newton methods and Jacobian-free Newton methods, treated in Sects. 4.1.2 and 4.1.3 respectively, can also be seen as inexact Newton methods. The general inexact Newton method is shown in Algorithm 4.2.

Algorithm 4.2 Inexact Newton Method

$i := 0$
given initial solution $\mathbf{x}_0$
while not converged **do**
  solve $-J(\mathbf{x}_i)\mathbf{s}_i = \mathbf{F}(\mathbf{x}_i)$ such that $\|\mathbf{r}_i\| \leq \eta_i \|\mathbf{F}(\mathbf{x}_i)\|$
  update iterate $\mathbf{x}_{i+1} := \mathbf{x}_i + \mathbf{s}_i$
  $i := i + 1$
end while

The convergence behaviour of the method strongly depends on the choice of the forcing terms. Convergence results derived in [1] are summarised in Table 4.1. In Chap. 5 we present theoretical results on local convergence for inexact Newton methods, proving that for properly chosen forcing terms the local convergence factor is arbitrarily close to η_i in each iteration. This result is reflected by the final row of Table 4.1, where $\alpha > 0$ can be chosen arbitrarily small. The specific conditions under which these convergence results hold can be found in [1] and Chap. 5 respectively.

If a forcing term is chosen too small, then the nonlinear error generally is reduced much less than the linear error in that iteration. This is called oversolving. In general, the closer the current iterate is to the solution, the smaller the forcing terms can be chosen without oversolving. Over the years, a lot of effort has been invested in finding good strategies for choosing the forcing terms, see for instance [2, 3].

Table 4.1 Local convergence for inexact Newton methods

Forcing terms	Local convergence
$\eta_i < 1$	Linear
$\limsup_{i\to\infty} \eta_i = 0$	Superlinear
$\limsup_{i\to\infty} \frac{\eta_i}{\|\mathbf{F}_i\|^p} < \infty,\ p \in (0,1)$	Order at least $1+p$
Properly chosen $\eta_i < 1$	Factor $(1+\alpha)\,\eta_i$

4.1.2 Approximate Jacobian Newton

The Jacobian of the function $\mathbf{F}(\mathbf{x})$ is not always available in practice. For example, it is possible that $\mathbf{F}(\mathbf{x})$ can be evaluated in any point by some method, but no analytical formulation is known. Then it is impossible to calculate the derivatives analytically. Or, if an analytical form is available, calculating the derivatives may simply be too computationally expensive.

In such cases, the Newton method may be used with appropriate approximations of the Jacobian matrices. The most widely used Jacobian matrix approximation is based on finite differences:

$$J_{ij}(\mathbf{x}) = \frac{\partial F_i}{\partial x_j}(\mathbf{x}) \approx \frac{F_i(\mathbf{x} + \delta \mathbf{e}_j) - F_i(\mathbf{x})}{\delta}, \tag{4.10}$$

where $\mathbf{e}_j$ is the vector with element j equal to 1, and all other elements equal to 0. For small enough δ, this is a good approximation of the derivative.

4.1.3 Jacobian-Free Newton

In some Newton–Raphson procedures the use of an explicit Jacobian matrix can be avoided. If done so, the method is called a Jacobian-free Newton method. Such a method may be used if the nonlinear problem is too large for the Jacobian to be stored in memory explicitly. Jacobian-free Newton methods can also be used as an alternative to approximate Jacobian Newton methods, if no analytical formulation of $\mathbf{F}(\mathbf{x})$ is known, or if the Jacobian is too computationally expensive to calculate.

Consider Newton–Krylov methods, where the Krylov solver only uses the Jacobian in matrix-vector products of the form $J(\mathbf{x})\mathbf{v}$. These products can be approximated by the directional finite difference scheme

$$J(\mathbf{x})\mathbf{v} \approx \frac{\mathbf{F}(\mathbf{x} + \delta \mathbf{v}) - \mathbf{F}(\mathbf{x})}{\delta}, \tag{4.11}$$

removing the need to store the Jacobian matrix explicitly.

For more information see [4], and the references therein.

4.2 Newton–Raphson with Global Convergence

Line search and trust region methods are iterative processes that can be used to find a local minimum in unconstrained optimisation. Both methods have global convergence to such a minimiser.

Unconstrained optimisation techniques can also be used to find roots of $\|\mathbf{F}\|$, which are the solutions of the nonlinear problem (4.2). Since line search and trust region methods ensure global convergence to a local minimum of $\|\mathbf{F}\|$, if all such minima are roots of $\mathbf{F}$, then these methods have global convergence to a solution of the nonlinear problem. However, if there is a local minimum that is not a root of $\|\mathbf{F}\|$, then the algorithm may terminate without finding a solution. In this case, the method is usually restarted from a different initial iterate, in the hope of finding a different local minimum that is a solution of the nonlinear system.

Near the solution, line search and trust region methods generally converge much slower than the Newton–Raphson method, but they can be used in conjunction with the Newton process to improve convergence farther from the solution. Both line search and trust region methods use their own criterion that has to be satisfied by the update vector. Whenever the Newton step satisfies this criterion then it is used to update the iterate normally. If the criterion is not satisfied, an alternative update vector is calculated that does satisfy the criterion, as detailed below.

4.2.1 Line Search

The idea behind augmenting the Newton–Raphson method with line search is simple. Instead of updating the iterate $\mathbf{x}_i$ with the Newton step $\mathbf{s}_i^N$, it is updated with some vector $\mathbf{s}_i = \lambda_i \mathbf{s}_i^N$ along the Newton step direction, i.e.,

$$\mathbf{x}_{i+1} = \mathbf{x}_i + \lambda_i \mathbf{s}_i^N. \tag{4.12}$$

Ideally, λ_i is chosen such that the nonlinear residual norm $\|\mathbf{F}(\mathbf{x}_i) + \lambda_i \mathbf{s}_i^N\|$ is minimised over all λ_i. Below, a strategy is outlined for finding a good value for λ_i, starting with the introduction of a convenient mathematical description of the problem. Note that $\mathbf{F}(\mathbf{x}_i) \neq \mathbf{0}$, as otherwise the nonlinear problem has already been solved with solution $\mathbf{x}_i$. In the remainder of this section, the iteration index i is dropped for readability.

Define the positive function

$$f(\mathbf{x}) = \frac{1}{2}\|\mathbf{F}(\mathbf{x})\|^2 = \frac{1}{2}\mathbf{F}(\mathbf{x})^T \mathbf{F}(\mathbf{x}), \tag{4.13}$$

and note that

$$\nabla f(\mathbf{x}) = J(\mathbf{x})^T \mathbf{F}(\mathbf{x}). \tag{4.14}$$

A vector $\mathbf{s}$ is called a descent direction of f in $\mathbf{x}$, if

$$\nabla f(\mathbf{x})^T \mathbf{s} < 0. \tag{4.15}$$

The Newton direction $\mathbf{s}^N = -J(\mathbf{x})^{-1}\mathbf{F}(\mathbf{x})$ is a descent direction, since

$$\nabla f(\mathbf{x})^T \mathbf{s}^N = -\mathbf{F}(\mathbf{x})^T J(\mathbf{x}) J(\mathbf{x})^{-1}\mathbf{F}(\mathbf{x}) = -\|\mathbf{F}(\mathbf{x})\|^2 < 0. \tag{4.16}$$

Now define the nonnegative function

$$g(\lambda) = f(\mathbf{x} + \lambda \mathbf{s}^N) = \frac{1}{2}\mathbf{F}(\mathbf{x} + \lambda \mathbf{s}^N)^T \mathbf{F}(\mathbf{x} + \lambda \mathbf{s}^N). \tag{4.17}$$

A minimiser of g also minimises $\|\mathbf{F}(\mathbf{x} + \lambda \mathbf{s}^N)\|$. Thus the best choice for λ is

$$\hat{\lambda} = \arg\min_{\lambda} g(\lambda). \tag{4.18}$$

It is generally not possible to solve the minimisation problem (4.18) analytically, but there are plenty methods to find a numerical approximation of $\hat{\lambda}$. In practice, a rough estimate suffices.

The decrease of f is regarded as sufficient, if λ satisfies the Armijo rule [5]

$$f(\mathbf{x} + \lambda \mathbf{s}^N) \leq f(\mathbf{x}) + \alpha\lambda \, \nabla f(\mathbf{x})^T \mathbf{s}^N, \tag{4.19}$$

where $\alpha \in (0, 1)$. A typical choice that often yields good results is $\alpha = 10^{-4}$. Note that for the Newton direction, we can also write the Armijo rule (4.19) as

$$\|\mathbf{F}(\mathbf{x} + \lambda \mathbf{s}^N)\|^2 \leq (1 - 2\alpha\lambda)\, \|\mathbf{F}(\mathbf{x})\|^2. \tag{4.20}$$

The common method to find a satisfactory value for λ, is to start with $\lambda^0 = 1$, and backtrack—as long as relation (4.19) is not satisfied—by setting

$$\lambda^{k+1} = \rho^k \lambda^k, \quad \rho^k \in [0.1, 0.5]\,. \tag{4.21}$$

The interval restriction on ρ^k is called safeguarding.

Since $\mathbf{s}^N$ is a descent direction, at some point the Armijo rule should be satisfied. The reduction factor ρ^k for λ^k, is chosen such that

$$\rho^k = \arg\min_{\rho^k \in [0.1, 0.5]} h(\rho^k \lambda^k), \tag{4.22}$$

where h is a quadratic polynomial model of f. This model h is constructed to be a parabola through either the values $g(0)$, $g'(0)$, and $g(\lambda^k)$, or the values $g(0)$, $g(\lambda^{k-1})$, and $g(\lambda^k)$. Note that for the Newton direction

$$g'(0) = \nabla f(\mathbf{x})^T \mathbf{s}^N = -\|\mathbf{F}(\mathbf{x})\|^2. \tag{4.23}$$

Further note that the second model can only be used from the second iteration, and λ^1 has to be chosen without the use of the model, for example by setting $\lambda^1 = 0.5$.

For more information on line search methods see for example [6]. For line search applied to inexact Newton–Krylov methods, see [7].

4.2.2 Trust Regions

Trust region methods define a region around the current iterate $\mathbf{x}_i$ that is trusted, and require the update step $\mathbf{s}_i$ to be such that the new iterate $\mathbf{x}_{i+1} = \mathbf{x}_i + \mathbf{s}_i$ lies within this trusted region. In this section the iteration index i is again dropped for readability.

Assume the trust region to be a hypersphere, i.e.,

$$\|\mathbf{s}\| \leq \delta. \tag{4.24}$$

The goal is to find the best possible update within the trust region.

Finding the update that minimises $\|\mathbf{F}\|$ within the trust region may be as hard as solving the nonlinear problem itself. Instead, the method searches for an update that satisfies

$$\min_{\|\mathbf{s}\| \leq \delta} q(\mathbf{s}), \tag{4.25}$$

with $q(\mathbf{s})$ the quadratic model of $\mathbf{F}(\mathbf{x} + \mathbf{s})$ given by

$$q(\mathbf{s}) = \frac{1}{2}\|\mathbf{r}\|^2 = \frac{1}{2}\|\mathbf{F} + J\mathbf{s}\|^2 = \frac{1}{2}\mathbf{F}^T\mathbf{F} + \left(J^T\mathbf{F}\right)^T \mathbf{s} + \frac{1}{2}\mathbf{s}^T J^T J\mathbf{s}, \tag{4.26}$$

where $\mathbf{F}$ and J are short for $\mathbf{F}(\mathbf{x})$ and $J(\mathbf{x})$ respectively.

The global minimum of the quadratic model $q(\mathbf{s})$, is attained at the Newton step $\mathbf{s}^N = -J(\mathbf{x})^{-1}\mathbf{F}(\mathbf{x})$, with $q(\mathbf{s}^N) = 0$. Thus, if the Newton step is within the trust region, i.e., if $\|\mathbf{s}^N\| \leq \delta$, then the current iterate is updated with the Newton step. However, if the Newton step is outside the trust region, it is not a valid update step.

It has been proven that problem (4.25) is solved by

$$\mathbf{s}(\mu) = \left(J(\mathbf{x})^T J(\mathbf{x}) + \mu I\right)^{-1} J(\mathbf{x})^T \mathbf{F}(\mathbf{x}), \tag{4.27}$$

for the unique μ for which $\|\mathbf{s}(\mu)\| = \delta$. See for example [6, Lemma 6.4.1], or [8, Theorem 7.2.1].

Finding this update vector $\mathbf{s}(\mu)$ is difficult, but there are fast methods to get a useful estimate, such as the hook step and the (double) dogleg step. The hook step method uses an iterative process to calculate update steps $\mathbf{s}(\mu)$ until $\|\mathbf{s}(\mu)\| \approx \delta$. Dogleg steps are calculated by constructing a piecewise linear approximation of the curve $\mathbf{s}(\mu)$, and setting the update step $\mathbf{s}$ to be the point where this approximation curve intersects the trust region boundary.

An essential part of making trust region methods work, is using suitable trust regions. Each time a new iterate is calculated it has to be decided if it is acceptable, and the size of the trust region has to be adjusted accordingly.

For an extensive treatment of trust regions methods see [8]. Further information on the application of trust region methods to inexact Newton–Krylov methods can be found in [7].

References

1. Dembo, R.S., Eisenstat, S.C., Steihaug, T.: Inexact Newton methods. SIAM J. Numer. Anal. **19**(2), 400–408 (1982)
2. Dembo, R.S., Steihaug, T.: Truncated-Newton algorithms for large-scale unconstrained optimization. Math. Program. **26**, 190–212 (1983)
3. Eisenstat, S.C., Walker, H.F.: Choosing the forcing terms in an inexact Newton method. SIAM J. Sci. Comput. **17**(1), 16–32 (1996)
4. Knoll, D.A., Keyes, D.E.: Jacobian-free Newton-Krylov methods: a survey of approaches and applications. J. Comput. Phys. **193**, 357–397 (2004)
5. Armijo, L.: Minimization of functions having lipschitz continuous irst partial derivatives. Pacific J. Math. **16**(1), 1–3 (1966)
6. Dennis Jr, J.E., Schnabel, R.B.: Numerical Methods for Unconstrained Optimization and Nonlinear Equations. Prentice Hall, New Jersey (1983)
7. Brown, P.N., Saad, Y.: Hybrid Krylov methods for nonlinear systems of equations. SIAM J. Sci. Stat. Comput. **11**(3), 450–481 (1990)
8. Conn, A.R., Gould, N.I.M., Toint, P.L.: Trust-Region Methods. SIAM, Philadelphia (2000)

Chapter 5
Convergence Theory

The Newton–Raphson method is usually the method of choice when solving systems of nonlinear equations. Good convergence properties reduce the number of Newton iterations needed to solve the problem, which is crucial for solving the problem in as little computational time as possible. However, the computational effort may not be the same in each Newton iteration, especially not for inexact Newton methods. Thus there is more to minimising the computational cost, than just minimising the number of Newton iterations.

A solid understanding of convergence behaviour is essential to the design and analysis of iterative methods. In this chapter we explore the convergence of inexact iterative methods in general, and inexact Newton methods in particular. A direct relationship between the convergence of inexact Newton methods and the forcing terms is presented, and the practical implications concerning computational effort are discussed and illustrated through numerical experiments.

5.1 Convergence of Inexact Iterative Methods

Assume an iterative method that, given current iterate $\mathbf{x}_i$, has some way to exactly determine a unique new iterate $\hat{\mathbf{x}}_{i+1}$. If instead an approximation $\mathbf{x}_{i+1}$ of the exact iterate $\hat{\mathbf{x}}_{i+1}$ is used to continue the process, we speak of an inexact iterative method. Inexact Newton methods (see Sect. 4.1.1) are examples of inexact iterative methods. Figure 5.1 illustrates a single step of an inexact iterative method.

Note that

$$\delta^c = \|\mathbf{x}_i - \hat{\mathbf{x}}_{i+1}\| > 0, \tag{5.1}$$

$$\delta^n = \|\mathbf{x}_{i+1} - \hat{\mathbf{x}}_{i+1}\| \geq 0, \tag{5.2}$$

$$\varepsilon^c = \|\mathbf{x}_i - \mathbf{x}^*\| > 0 \tag{5.3}$$

$$\varepsilon^n = \|\mathbf{x}_{i+1} - \mathbf{x}^*\|, \tag{5.4}$$

$$\hat{\varepsilon} = \|\hat{\mathbf{x}}_{i+1} - \mathbf{x}^*\| \geq 0. \tag{5.5}$$

R. Idema and D. J. P. Lahaye, *Computational Methods in Power System Analysis*,
Atlantis Studies in Scientific Computing in Electromagnetics,
DOI: 10.2991/978-94-6239-064-5_5, © Atlantis Press and the authors 2014

Fig. 5.1 Inexact iterative step

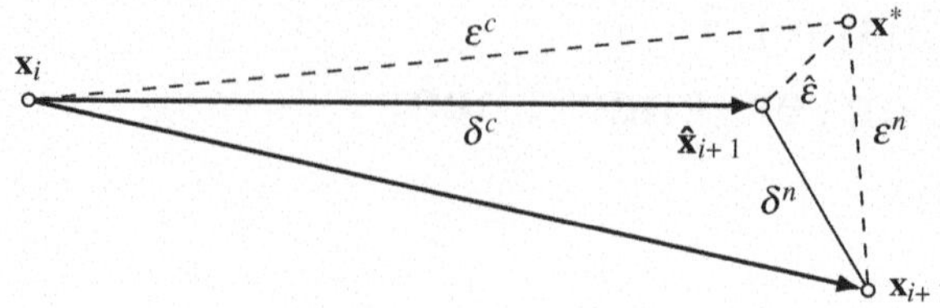

Define γ as the distance of the exact iterate $\hat{\mathbf{x}}_{i+1}$ to the solution, relative to the length δ^c of the exact update step, i.e.,

$$\gamma = \frac{\hat{\varepsilon}}{\delta^c} > 0. \tag{5.6}$$

The ratio $\frac{\varepsilon^n}{\varepsilon^c}$ is a measure for the improvement of the inexact iterate $\mathbf{x}_{i+1}$ over the current iterate $\mathbf{x}_i$, in terms of the distance to the solution $\mathbf{x}^*$. Likewise, the ratio $\frac{\delta^n}{\delta^c}$ is a measure for the improvement of the inexact iterate $\mathbf{x}_{i+1}$, in terms of the distance to the exact iterate $\hat{\mathbf{x}}_{i+1}$. As the solution is unknown, so is the ratio $\frac{\varepsilon^n}{\varepsilon^c}$. Assume, however, that some measure for the ratio $\frac{\delta^n}{\delta^c}$ is available, and that it can be controlled. For example, for an inexact Newton method the forcing terms η_i can be used to control $\frac{\delta^n}{\delta^c}$.

The aim is to have an improvement of the controllable error impose a similar improvement on the distance to the solution, i.e., to have

$$\frac{\varepsilon^n}{\varepsilon^c} \leq (1+\alpha)\frac{\delta^n}{\delta^c} \tag{5.7}$$

for some reasonably small $\alpha > 0$.

The worst case scenario can be identified as

$$\max \frac{\varepsilon^n}{\varepsilon^c} = \frac{\delta^n + \hat{\varepsilon}}{\left|\delta^c - \hat{\varepsilon}\right|} = \frac{\delta^n + \gamma\delta^c}{|1-\gamma|\,\delta^c} = \frac{1}{|1-\gamma|}\frac{\delta^n}{\delta^c} + \frac{\gamma}{|1-\gamma|}. \tag{5.8}$$

To guarantee that the inexact iterate $\mathbf{x}_{i+1}$ is an improvement over $\mathbf{x}_i$, using Eq. (5.8), it is required that

$$\frac{1}{|1-\gamma|}\frac{\delta^n}{\delta^c} + \frac{\gamma}{|1-\gamma|} < 1 \Leftrightarrow \frac{\delta^n}{\delta^c} + \gamma < |1-\gamma| \Leftrightarrow \frac{\delta^n}{\delta^c} < |1-\gamma| - \gamma. \tag{5.9}$$

If $\gamma \geq 1$ this would mean that $\frac{\delta^n}{\delta^c} < -1$, which is impossible. Therefore, to guarantee a reduction of the distance to the solution, it is required that

$$\frac{\delta^n}{\delta^c} < 1 - 2\gamma \Leftrightarrow 2\gamma < 1 - \frac{\delta^n}{\delta^c} \Leftrightarrow \gamma < \frac{1}{2} - \frac{1}{2}\frac{\delta^n}{\delta^c}. \tag{5.10}$$

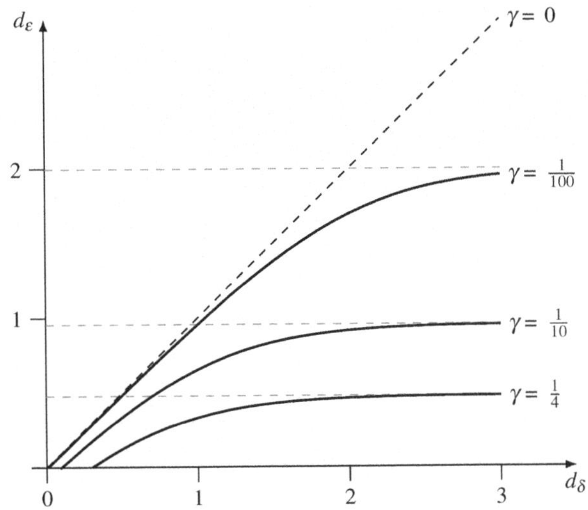

Fig. 5.2 Number of digits improvement

As a result, the absolute operators can be dropped from Eq. (5.8).

Note that if the iterative method converges to the solution superlinearly, then γ goes to 0 with the same rate of convergence. Thus, at some point in the iteration process Eq. (5.10) is guaranteed to hold. This is in particular the case for an inexact Newton method, if it converges, as convergence is quadratic once the iterate is close enough to the solution.

Figure 5.2 shows plots of Eq. (5.8) on a logarithmic scale for several values of γ. The horizontal axis shows the number of digits improvement in the distance to the exact iterate, and the vertical axis depicts the resulting minimum number of digits improvement in the distance to the solution, i.e.,

$$d_\delta = -\log \frac{\delta^n}{\delta^c} \quad \text{and} \quad d_\varepsilon = -\log \left(\max \frac{\varepsilon^n}{\varepsilon^c} \right). \tag{5.11}$$

For fixed d_δ, the smaller the value of γ, the better the resulting d_ε is. For $\gamma = \frac{1}{10}$, there is a significant start-up cost on d_δ before d_ε becomes positive, and a full digit improvement on the distance to the solution can never be guaranteed. Making more than a 2 digit improvement in the distance to the exact iterate results in a lot of effort with hardly any return at $\gamma = \frac{1}{10}$. However, when $\gamma = \frac{1}{100}$ there is hardly any start-up cost on d_δ any more, and the guaranteed improvement in the distance to the solution can be taken up to about 2 digits.

The above mentioned start-up cost can be derived from Eq. (5.10) to be $d_\delta = -\log(1 - 2\gamma)$, while the asymptotic value to which d_ε approaches is given by $d_\varepsilon = -\log\left(\frac{\gamma}{1-\gamma}\right) = \log\left(\frac{1}{\gamma} - 1\right)$, which is the improvement obtained when using the exact iterate.

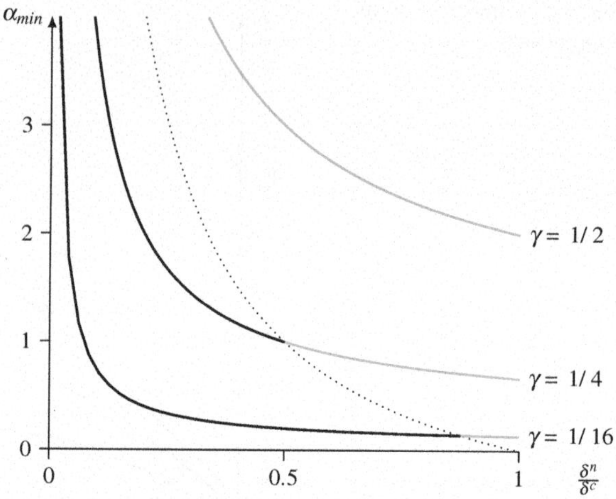

Fig. 5.3 Minimum required value of α

The value α, as introduced in Eq. (5.7), is a measure of how far the graph of d_ε deviates from the ideal $d_\varepsilon = d_\delta$, which is attained only in the fictitious case that $\gamma = 0$. Combining Eqs. (5.7) and (5.8), the minimum value of α that is needed for Eq. (5.7) to be guaranteed to hold can be investigated:

$$\frac{1}{1-\gamma}\frac{\delta^n}{\delta^c} + \frac{\gamma}{1-\gamma} = (1+\alpha_{min})\frac{\delta^n}{\delta^c} \Leftrightarrow \tag{5.12}$$

$$\frac{1}{1-\gamma} + \frac{\gamma}{1-\gamma}\left(\frac{\delta^n}{\delta^c}\right)^{-1} = (1+\alpha_{min}) \Leftrightarrow \tag{5.13}$$

$$\alpha_{min} = \frac{\gamma}{1-\gamma}\left[\left(\frac{\delta^n}{\delta^c}\right)^{-1} + 1\right]. \tag{5.14}$$

Figure 5.3 shows α_{min} as a function of $\frac{\delta^n}{\delta^c} \in [0, 1)$ for several values of γ. Left of the dotted line the Eq. (5.10) is satisfied, i.e., improvement of the distance to the solution is guaranteed, whereas right of the dotted line this is not the case.

For given γ, reducing $\frac{\delta^n}{\delta^c}$ increases α_{min}. Especially for small $\frac{\delta^n}{\delta^c}$, the value of α_{min} grows very rapidly. Thus, the closer the inexact iterate is brought to the exact iterate, the less the expected relative return in the distance to the solution is. For the inexact Newton method this translates into oversolving whenever the forcing term η_i is chosen too small.

Further, it is clear that if γ becomes smaller, then α_{min} is reduced also. If γ is small, $\frac{\delta^n}{\delta^c}$ can be made very small without compromising the return of investment on the distance to the solution. However, for γ nearing $\frac{1}{2}$, or more, no choice of $\frac{\delta^n}{\delta^c}$ can guarantee a similar improvement, if any, in the distance to the solution. Therefore, for such γ oversolving is inevitable.

Recall that if the iterative method converges superlinearly, then γ rapidly goes to 0 also. Thus, for such a method, $\frac{\delta^n}{\delta^c}$ can be made smaller and smaller in later iterations, without oversolving. In other words, for any choice of $\alpha > 0$ and $\frac{\delta^n}{\delta^c} \in [0, 1)$, there will be some point in the iteration process from which on forward Eq. (5.7) is satisfied.

When using an inexact Newton method $\frac{\delta^n}{\delta^c} = \frac{\|\mathbf{x}_{i+1}-\hat{\mathbf{x}}_{i+1}\|}{\|\mathbf{x}_i-\hat{\mathbf{x}}_{i+1}\|}$ is not actually known, but the relative residual error $\frac{\|\mathbf{r}_i\|}{\|\mathbf{F}(\mathbf{x}_i)\|} = \frac{\|J(\mathbf{x}_i)(\mathbf{x}_{i+1}-\hat{\mathbf{x}}_{i+1})\|}{\|J(\mathbf{x}_i)(\mathbf{x}_i-\hat{\mathbf{x}}_{i+1})\|}$, which is controlled by the forcing terms η_i, can be used as a measure for it. In the next section, this idea is used to proof a useful variation on Eq. (5.7) for inexact Newton methods.

5.2 Convergence of Inexact Newton Methods

Consider the nonlinear system of equations $\mathbf{F}(\mathbf{x}) = \mathbf{0}$, where:

- there is a solution $\mathbf{x}^*$ such that $\mathbf{F}(\mathbf{x}^*) = \mathbf{0}$,
- the Jacobian matrix J of $\mathbf{F}$ exists in a neighbourhood of $\mathbf{x}^*$,
- $J(\mathbf{x}^*)$ is continuous and nonsingular.

In this section, theory is presented that relates the convergence of the inexact Newton method, for a problem of the above form, directly to the chosen forcing terms. The following theorem is a variation on the inexact Newton convergence theorem presented in [1, Thm. 2.3].

Theorem 5.1. *Let $\eta_i \in (0, 1)$ and choose $\alpha > 0$ such that $(1+\alpha)\,\eta_i < 1$. Then there exists an $\varepsilon > 0$ such that, if $\|\mathbf{x}_0 - \mathbf{x}^*\| < \varepsilon$, the sequence of inexact Newton iterates $\mathbf{x}_i$ converges to $\mathbf{x}^*$, with*

$$\|J(\mathbf{x}^*)\left(\mathbf{x}_{i+1} - \mathbf{x}^*\right)\| < (1+\alpha)\,\eta_i\|J(\mathbf{x}^*)\left(\mathbf{x}_i - \mathbf{x}^*\right)\|. \tag{5.15}$$

Proof. Define

$$\mu = \max[\|J(\mathbf{x}^*)\|, \|J(\mathbf{x}^*)^{-1}\|] \geq 1. \tag{5.16}$$

Recall that $J(\mathbf{x}^*)$ is nonsingular. Thus μ is well-defined and we can write

$$\frac{1}{\mu}\|\mathbf{y}\| \leq \|J(\mathbf{x}^*)\mathbf{y}\| \leq \mu\|\mathbf{y}\|. \tag{5.17}$$

Note that $\mu \geq 1$ because the induced matrix norm is submultiplicative.

Let

$$\gamma \in \left(0, \frac{\alpha\eta_i}{5\mu}\right) \tag{5.18}$$

and choose $\varepsilon > 0$ sufficiently small such that if $\|\mathbf{y} - \mathbf{x}^*\| \leq \mu^2\varepsilon$ then

$$\|J(\mathbf{y}) - J(\mathbf{x}^*)\| \le \gamma, \tag{5.19}$$

$$\|J(\mathbf{y})^{-1} - J(\mathbf{x}^*)^{-1}\| \le \gamma, \tag{5.20}$$

$$\|\mathbf{F}(\mathbf{y}) - \mathbf{F}(\mathbf{x}^*) - J(\mathbf{x}^*)\left(\mathbf{y} - \mathbf{x}^*\right)\| \le \gamma \|\mathbf{y} - \mathbf{x}^*\|. \tag{5.21}$$

That such an ε exists follows from [2, Thm. 2.3.3 & 3.1.5].

First we show that if $\|\mathbf{x}_i - \mathbf{x}^*\| < \mu^2\varepsilon$, then Eq. (5.15) holds.

Write

$$J(\mathbf{x}^*)\left(\mathbf{x}_{i+1} - \mathbf{x}^*\right) = \left[I + J(\mathbf{x}^*)\left(J(\mathbf{x}_i)^{-1} - J(\mathbf{x}^*)^{-1}\right)\right] \cdot [\mathbf{r}_i + \left(J(\mathbf{x}_i) - J(\mathbf{x}^*)\right)\left(\mathbf{x}_i - \mathbf{x}^*\right) - \left(\mathbf{F}(\mathbf{x}_i) - \mathbf{F}(\mathbf{x}^*) - J(\mathbf{x}^*)\left(\mathbf{x}_i - \mathbf{x}^*\right)\right)]. \tag{5.22}$$

Taking norms gives

$$\begin{aligned}\|J(\mathbf{x}^*)\left(\mathbf{x}_{i+1} - \mathbf{x}^*\right)\| &\le \left[1 + \|J(\mathbf{x}^*)\|\|J(\mathbf{x}_i)^{-1} - J(\mathbf{x}^*)^{-1}\|\right] \cdot [\|\mathbf{r}_i\| + \\ &\quad \|J(\mathbf{x}_i) - J(\mathbf{x}^*)\|\|\mathbf{x}_i - \mathbf{x}^*\| + \|\mathbf{F}(\mathbf{x}_i) - \mathbf{F}(\mathbf{x}^*) - J(\mathbf{x}^*)\left(\mathbf{x}_i - \mathbf{x}^*\right)\|], \\ &\le \left[1 + \mu\gamma\right] \cdot \left[\|\mathbf{r}_i\| + \gamma\|\mathbf{x}_i - \mathbf{x}^*\| + \gamma\|\mathbf{x}_i - \mathbf{x}^*\|\right], \\ &\le \left[1 + \mu\gamma\right] \cdot \left[\eta_i\|\mathbf{F}(\mathbf{x}_i)\| + 2\gamma\|\mathbf{x}_i - \mathbf{x}^*\|\right].\end{aligned} \tag{5.23}$$

Here the definitions of η_i (Eq. 4.8) and μ (Eq. 5.16) were used, together with Eqs. (5.19–5.21).

Further write, using that by definition $\mathbf{F}(\mathbf{x}^*) = \mathbf{0}$,

$$\mathbf{F}(\mathbf{x}_i) = \left[J(\mathbf{x}^*)\left(\mathbf{x}_i - \mathbf{x}^*\right)\right] + \left[\mathbf{F}(\mathbf{x}_i) - \mathbf{F}(\mathbf{x}^*) - J(\mathbf{x}^*)\left(\mathbf{x}_i - \mathbf{x}^*\right)\right]. \tag{5.24}$$

Again taking norms gives

$$\begin{aligned}\|\mathbf{F}(\mathbf{x}_i)\| &\le \|J(\mathbf{x}^*)\left(\mathbf{x}_i - \mathbf{x}^*\right)\| + \|\mathbf{F}(\mathbf{x}_i) - \mathbf{F}(\mathbf{x}^*) - J(\mathbf{x}^*)\left(\mathbf{x}_i - \mathbf{x}^*\right)\| \\ &\le \|J(\mathbf{x}^*)\left(\mathbf{x}_i - \mathbf{x}^*\right)\| + \gamma\|\mathbf{x}_i - \mathbf{x}^*\|.\end{aligned} \tag{5.25}$$

Substituting Eq. (5.25) into (5.23) then leads to

$$\begin{aligned}&\|J(\mathbf{x}^*)\left(\mathbf{x}_{i+1} - \mathbf{x}^*\right)\| \\ &\le (1 + \mu\gamma)\left[\eta_i\left(\|J(\mathbf{x}^*)\left(\mathbf{x}_i - \mathbf{x}^*\right)\| + \gamma\|\mathbf{x}_i - \mathbf{x}^*\|\right) + 2\gamma\|\mathbf{x}_i - \mathbf{x}^*\|\right] \\ &\le (1 + \mu\gamma)\left[\eta_i(1 + \mu\gamma) + 2\mu\gamma\right]\|J(\mathbf{x}^*)\left(\mathbf{x}_i - \mathbf{x}^*\right)\|.\end{aligned} \tag{5.26}$$

Here Eq. (5.17) was used to write $\|\mathbf{x}_i - \mathbf{x}^*\| \le \mu\|J(\mathbf{x}^*)\left(\mathbf{x}_i - \mathbf{x}^*\right)\|$.

Finally, using that $\gamma \in \left(0, \frac{\alpha\eta_i}{5\mu}\right)$, and that both $\eta_i < 1$ and $\alpha\eta_i < 1$—the latter being a result from the requirement that $(1 + \alpha)\,\eta_i < 1$—gives

$$\begin{aligned}
(1+\mu\gamma)\left[\eta_i(1+\mu\gamma)+2\mu\gamma\right] &\leq \left(1+\frac{\alpha\eta_i}{5}\right)\left[\eta_i\left(1+\frac{\alpha\eta_i}{5}\right)+\frac{2\alpha\eta_i}{5}\right] \\
&= \left[\left(1+\frac{\alpha\eta_i}{5}\right)^2+\left(1+\frac{\alpha\eta_i}{5}\right)\frac{2\alpha}{5}\right]\eta_i \\
&= \left[1+\frac{2\alpha\eta_i}{5}+\frac{\alpha^2\eta_i^2}{25}+\frac{2\alpha}{5}+\frac{2\alpha^2\eta_i}{25}\right]\eta_i \\
&< \left[1+\frac{2\alpha}{5}+\frac{\alpha}{25}+\frac{2\alpha}{5}+\frac{2\alpha}{25}\right]\eta_i \\
&< (1+\alpha)\,\eta_i. \qquad (5.27)
\end{aligned}$$

Equation (5.15) follows by substituting Eq. (5.27) into Eq. (5.26).

Given that Eq. (5.15) holds if $\|\mathbf{x}_i - \mathbf{x}^*\| < \mu^2\varepsilon$, we proceed to prove Theorem 5.1 by induction.

For the base case

$$\|\mathbf{x}_0 - \mathbf{x}^*\| < \varepsilon \leq \mu^2\varepsilon. \qquad (5.28)$$

Thus Eq. (5.15) holds for $i = 0$.

The induction hypothesis that Eq. (5.15) holds for $i = 0, \ldots, k-1$ then gives

$$\begin{aligned}
\|\mathbf{x}_k - \mathbf{x}^*\| &\leq \mu\|J(\mathbf{x}^*)\left(\mathbf{x}_k - \mathbf{x}^*\right)\| \\
&< \mu\,(1+\alpha)^k\,\eta_{k-1}\cdots\eta_0\|J(\mathbf{x}^*)\left(\mathbf{x}_0 - \mathbf{x}^*\right)\| \\
&< \mu\|J(\mathbf{x}^*)\left(\mathbf{x}_0 - \mathbf{x}^*\right)\| \\
&\leq \mu^2\|\mathbf{x}_0 - \mathbf{x}^*\| \\
&< \mu^2\varepsilon. \qquad (5.29)
\end{aligned}$$

Thus Eq. (5.15) also holds for $i = k$, completing the proof. □

In words, Theorem 5.1 states that for an arbitrarily small $\alpha > 0$, and any choice of forcing terms $\eta_i \in (0, 1)$, Eq. (5.15) holds if the current iterate is close enough to the solution.

Note that this does not mean that for a certain iterate $\mathbf{x}_i$, one can choose α and η_i arbitrarily small and expect Eq. (5.15) to hold, as ε depends on the choice of α and η_i. On the contrary, a given iterate $\mathbf{x}_i$—close enough to the solution to guarantee convergence—imposes the restriction that, for Theorem 5.1 to hold, the forcing terms η_i cannot be chosen too small. Recall that it was already shown in Sect. 5.1 that choosing η_i too small leads to oversolving.

If we define oversolving as using forcing terms η_i that are too small for a certain iterate $\mathbf{x}_i$, in the context of Theorem 5.1, then the theorem can be characterised by saying that a convergence factor $(1+\alpha)\,\eta_i$ is attained if η_i is chosen such that there is no oversolving. Using Eq. (5.18), $\eta_i > \frac{5\mu\gamma}{\alpha}$ can then be seen as a theoretical bound on the forcing terms that guards against oversolving.

Corollary 5.1. *Let $\eta_i \in (0,1)$ and choose $\alpha > 0$ such that $(1+\alpha)\,\eta_i < 1$. Then there exists an $\varepsilon > 0$ such that, if $\|\mathbf{x}_0 - \mathbf{x}^*\| < \varepsilon$, the sequence of inexact Newton iterates $\mathbf{x}_i$ converges to $\mathbf{x}^*$, with*

$$\|J(\mathbf{x}^*)\left(\mathbf{x}_i - \mathbf{x}^*\right)\| < (1+\alpha)^i\, \eta_{i-1} \cdots \eta_0 \|J(\mathbf{x}^*)\left(\mathbf{x}_0 - \mathbf{x}^*\right)\|. \tag{5.30}$$

Proof. The stated follows readily from the repeated application of Theorem 5.1. □

A relation between the nonlinear residual norm $\|\mathbf{F}(\mathbf{x}_i)\|$ and the error norm $\|J(\mathbf{x}^*)\,(\mathbf{x}_i - \mathbf{x}^*)\|$ can be derived, within the neighbourhood of the solution where Theorem 5.1 holds.

Theorem 5.2. *Let $\eta_i \in (0,1)$ and choose $\alpha > 0$ such that $(1+\alpha)\,\eta_i < 1$. Then there exists an $\varepsilon > 0$ such that, if $\|\mathbf{x}_0 - \mathbf{x}^*\| < \varepsilon$, then*

$$\left(1 - \frac{\alpha\eta_i}{5}\right) \|J(\mathbf{x}^*)\left(\mathbf{x}_i - \mathbf{x}^*\right)\| < \|\mathbf{F}(\mathbf{x}_i)\| < \left(1 + \frac{\alpha\eta_i}{5}\right) \|J(\mathbf{x}^*)\left(\mathbf{x}_i - \mathbf{x}^*\right)\|. \tag{5.31}$$

Proof. Using that $\mathbf{F}(\mathbf{x}^*) = \mathbf{0}$ by definition, again write

$$\mathbf{F}(\mathbf{x}_i) = \left[J(\mathbf{x}^*)\left(\mathbf{x}_i - \mathbf{x}^*\right)\right] + \left[\mathbf{F}(\mathbf{x}_i) - \mathbf{F}(\mathbf{x}^*) - J(\mathbf{x}^*)\left(\mathbf{x}_i - \mathbf{x}^*\right)\right]. \tag{5.32}$$

Taking norms, and using Eqs. (5.21) and (5.17), gives

$$\begin{aligned}
\|\mathbf{F}(\mathbf{x}_i)\| &\le \|J(\mathbf{x}^*)\left(\mathbf{x}_i - \mathbf{x}^*\right)\| + \|\mathbf{F}(\mathbf{x}_i) - \mathbf{F}(\mathbf{x}^*) - J(\mathbf{x}^*)\left(\mathbf{x}_i - \mathbf{x}^*\right)\| \\
&\le \|J(\mathbf{x}^*)\left(\mathbf{x}_i - \mathbf{x}^*\right)\| + \gamma\|\mathbf{x}_i - \mathbf{x}^*\| \\
&\le \|J(\mathbf{x}^*)\left(\mathbf{x}_i - \mathbf{x}^*\right)\| + \mu\gamma\|J(\mathbf{x}^*)\mathbf{x}_i - \mathbf{x}^*\| \\
&= (1+\mu\gamma)\,\|J(\mathbf{x}^*)\left(\mathbf{x}_i - \mathbf{x}^*\right)\|.
\end{aligned} \tag{5.33}$$

Similarly, it holds that

$$\begin{aligned}
\|\mathbf{F}(\mathbf{x}_i)\| &\ge \|J(\mathbf{x}^*)\left(\mathbf{x}_i - \mathbf{x}^*\right)\| - \|\mathbf{F}(\mathbf{x}_i) - \mathbf{F}(\mathbf{x}^*) - J(\mathbf{x}^*)\left(\mathbf{x}_i - \mathbf{x}^*\right)\| \\
&\ge \|J(\mathbf{x}^*)\left(\mathbf{x}_i - \mathbf{x}^*\right)\| - \gamma\|\mathbf{x}_i - \mathbf{x}^*\| \\
&\ge \|J(\mathbf{x}^*)\left(\mathbf{x}_i - \mathbf{x}^*\right)\| - \mu\gamma\|J(\mathbf{x}^*)\mathbf{x}_i - \mathbf{x}^*\| \\
&= (1-\mu\gamma)\,\|J(\mathbf{x}^*)\left(\mathbf{x}_i - \mathbf{x}^*\right)\|.
\end{aligned} \tag{5.34}$$

The theorem now follows from Eq. (5.18). □

For the nonlinear residual norm $\|\mathbf{F}(\mathbf{x}_i)\|$, a similar result can now be derived as was presented for the error norm $\|J(\mathbf{x}^*)\,(\mathbf{x}_i - \mathbf{x}^*)\|$ in Theorem 5.1.

Theorem 5.3. *Let $\eta_i \in (0,1)$ and choose $\alpha > 0$ such that $(1+2\alpha)\,\eta_i < 1$. Then there exists an $\varepsilon > 0$ such that, if $\|\mathbf{x}_0 - \mathbf{x}^*\| < \varepsilon$, the sequence $\|\mathbf{F}(\mathbf{x}_i)\|$ converges to 0, with*

$$\|\mathbf{F}(\mathbf{x}_{i+1})\| < (1+2\alpha)\,\eta_i\|\mathbf{F}(\mathbf{x}_i)\|. \tag{5.35}$$

Proof. Note that the conditions imposed in Theorem 5.3, are such that Theorems 5.1 and 5.2 hold. Define μ and γ again as in Theorem 5.1.

Using Eq. (5.33), Theorem 5.1, and Eq. (5.34), write

$$\begin{aligned}\|\mathbf{F}(\mathbf{x}_{i+1})\| &\leq (1+\mu\gamma)\,\|J(\mathbf{x}^*)\left(\mathbf{x}_{i+1}-\mathbf{x}^*\right)\| \\ &< (1+\mu\gamma)\,(1+\alpha)\,\eta_i\|J(\mathbf{x}^*)\left(\mathbf{x}_i-\mathbf{x}^*\right)\| \\ &\leq \frac{(1+\mu\gamma)}{(1-\mu\gamma)}\,(1+\alpha)\,\eta_i\|\mathbf{F}(\mathbf{x}_i)\|. \end{aligned} \tag{5.36}$$

Further, using Eq. (5.18), write

$$\frac{1+\mu\gamma}{1-\mu\gamma} < \frac{1+\frac{\alpha\eta_i}{5}}{1-\frac{\alpha\eta_i}{5}} = \frac{1-\frac{\alpha\eta_i}{5}+\frac{2}{5}\alpha\eta_i}{1-\frac{\alpha\eta_i}{5}} = 1+\frac{\frac{2}{5}\alpha\eta_i}{1-\frac{\alpha\eta_i}{5}} < 1+\frac{\frac{2}{5}\alpha\eta_i}{\frac{4}{5}} = 1+\frac{\alpha\eta_i}{2}.$$

Finally, using that both $\eta_i < 1$ and $2\alpha\eta_i < 1$—the latter being a result from the requirement that $(1+2\alpha)\,\eta_i < 1$—gives

$$\frac{1+\mu\gamma}{1-\mu\gamma}\,(1+\alpha) < \left(1+\frac{\alpha\eta_i}{2}\right)(1+\alpha) = 1+\left(1+\frac{\eta_i}{2}\right)\alpha+\frac{1}{2}\eta_i\alpha^2 < 1+2\alpha$$

Substitution into Eq. (5.36) completes the proof. □

Theorem 5.3 shows that the nonlinear residual norm $\|\mathbf{F}(\mathbf{x}_i)\|$ converges at similar rate as the error norm $\|J(\mathbf{x}^*)\,(\mathbf{x}_i-\mathbf{x}^*)\|$. This is important, because Newton methods use $\|\mathbf{F}(\mathbf{x}_i)\|$ to measure convergence of the iterate to the solution.

5.2.1 *Linear Convergence*

The inexact Newton method uses some iterative process to solve the linear Jacobian system $J(\mathbf{x}_i)\mathbf{s}_i = -\mathbf{F}(\mathbf{x}_i)$ up to accuracy $\|J(\mathbf{x}_i)\mathbf{s}_i+\mathbf{F}(\mathbf{x}_i)\| \leq \eta_i\|\mathbf{F}(\mathbf{x}_i)\|$ in each Newton iteration. In many practical applications, the convergence of the iterative linear solver turns out to be approximately linear. That is, for some convergence rate $\beta > 0$

$$\|\mathbf{r}_i^k\| \approx 10^{-\beta k}\|\mathbf{F}(\mathbf{x}_i)\|, \tag{5.37}$$

where $\mathbf{r}_i^k = \mathbf{F}(\mathbf{x}_i)+J(\mathbf{x}_i)\mathbf{s}_i^k$ is the linear residual after k iterations of the linear solver in Newton iteration i.

Suppose that the linear solver indeed converges linearly, with the same rate of convergence β in each Newton iteration. Let K_i be the number of linear iterations performed in Newton iteration i, i.e., K_i is minimum integer such that $10^{-\beta K_i} \leq \eta_i$. Further, let $N_i = \sum_{j=0}^{i-1} K_j$ be the total number of linear iterations performed up to the start of Newton iteration i. Then, using Corollary 5.1,

$$\begin{aligned}\|J(\mathbf{x}^*)\left(\mathbf{x}_i - \mathbf{x}^*\right)\| &< (1+\alpha)^i\, \eta_{i-1}\cdots\eta_0\|J(\mathbf{x}^*)\left(\mathbf{x}_0 - \mathbf{x}^*\right)\| \\ &= (1+\alpha)^i\, 10^{-\beta N_i}\|J(\mathbf{x}^*)\left(\mathbf{x}_0 - \mathbf{x}^*\right)\|. \end{aligned} \tag{5.38}$$

Thus, if the linear solver converges approximately linearly, with similar rate of convergence in each Newton iteration, and the forcing terms are such that there is no oversolving, and if α can be chosen small enough, i.e., the initial iterate is close enough to the solution, then the inexact Newton method will converge approximately linearly in the total number of linear iterations.

Note that this result is independent of the rate of convergence in the Newton iterations. If the forcing terms are chosen constant, the method will converge linearly in the number of Newton iterations, and linearly in the total number of linear iterations performed throughout those Newton iterations. If the forcing terms η_i are chosen properly, the method will converge quadratically in the Newton iterations, while converging linearly in the linear iterations. The amount of Newton iterations needed in these two scenarios may differ greatly, but the total amount of linear iterations should be approximately equal.

5.3 Numerical Experiments

Both the classical Newton–Raphson convergence theory [2, 3], and the inexact Newton convergence theory by Dembo et al. [1], require the current iterate to be close enough to the solution. What exactly is "close enough" depends on the problem, and is in practice generally too difficult to calculate. However, decades of practice have shown that the theoretical convergence is reached within a few Newton steps for most problems, if needed using globalisation methods as described in Sect. 4.2. Thus the theory is not just of theoretical, but also of practical importance.

In this section, numerical experiments are presented to illustrate the practical merit of Theorem 5.1, despite the elusive requirement that the current iterate has to be close enough to the solution. Moreover, instead of convergence relation (5.15), an idealised version is tested, in which the error norm is changed to the 2-norm and α is neglected:

$$\|\mathbf{x}_{i+1} - \mathbf{x}^*\| < \eta_i\|\mathbf{x}_i - \mathbf{x}^*\|. \tag{5.39}$$

If relation (5.39) is satisfied, that means that any improvement of the linear residual norm in a certain Newton iteration, improves the error in the nonlinear iterate by an equal factor.

The experiments in this section are performed on a power flow problem. The power flow problem, and how to solve it, is treated in Chaps. 6–8. The actual test case used is the uctew032 power flow problem (see Chap. 11). The resulting nonlinear system has approximately 256k equations, and the Jacobian matrix has around 2M nonzeros. The linear Jacobian systems are solved using GMRES, preconditioned with a high quality ILU factorisation of the Jacobian.

In Figs. 5.4, 5.5, 5.6, the results are presented for different amounts of GMRES iterations per Newton iteration. In all cases, two Newton steps with just a single GMRES iteration were performed at the start, but omitted from the figure. In each figure the solid line represents the actual error norm $\|\mathbf{x}_i - \mathbf{x}^*\|$, while the dashed line depicts the expected error norm following the idealised theoretical relation (5.39).

Figure 5.4 has a distribution of GMRES iterations that leads to a fast solution of the problem. The practical convergence nicely follows the theory. This suggests that the two initial Newton iterations with a single GMRES iteration each, lead to an iterate $\mathbf{x}_2$ that is close enough to the solution to use the chosen forcing terms η_i without oversolving. Note that $\mathbf{x}_2$ is in actuality still very far from the solution, and that it is unlikely that it satisfies the theoretical bound on the proximity to the solution required in Theorem 5.1.

Figure 5.5 shows the convergence for a more exotic distribution of GMRES iterations per Newton iteration, illustrating that practice can also follow theory nicely for such a scenario.

Figure 5.6 illustrates the impact of oversolving. Practical convergence is nowhere near the idealised theory because extra GMRES iterations are performed that do not further improve the Newton error. In terms of Theorem 5.1 this means that the iterates $\mathbf{x}_i$ are not close enough to the solution to be able to use forcing terms η_i as small as they were effectively chosen in this example.

In Fig. 5.7, the convergence in the number of Newton iterations is compared with the convergence in the number of GMRES iterations. For the uctew032 test case, the convergence of the GMRES solves is approximately linear, with similar rate of

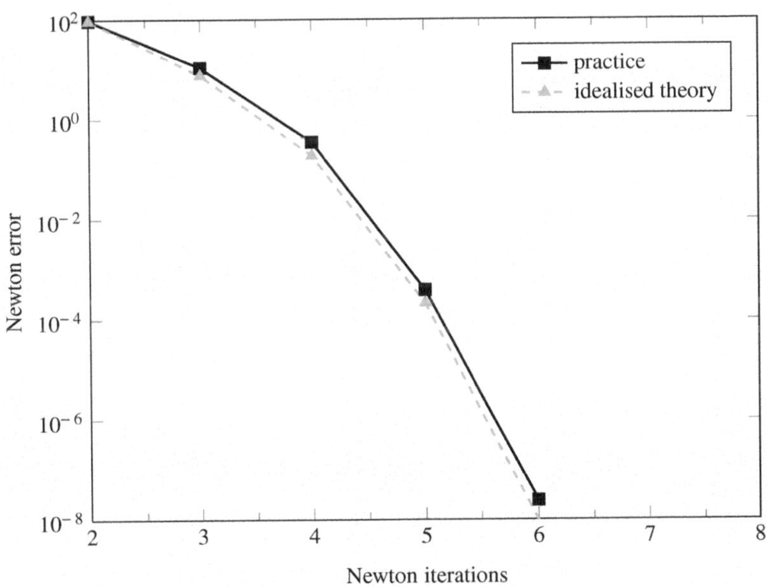

Fig. 5.4 GMRES iteration distribution 1, 1, 4, 6, 10, 14

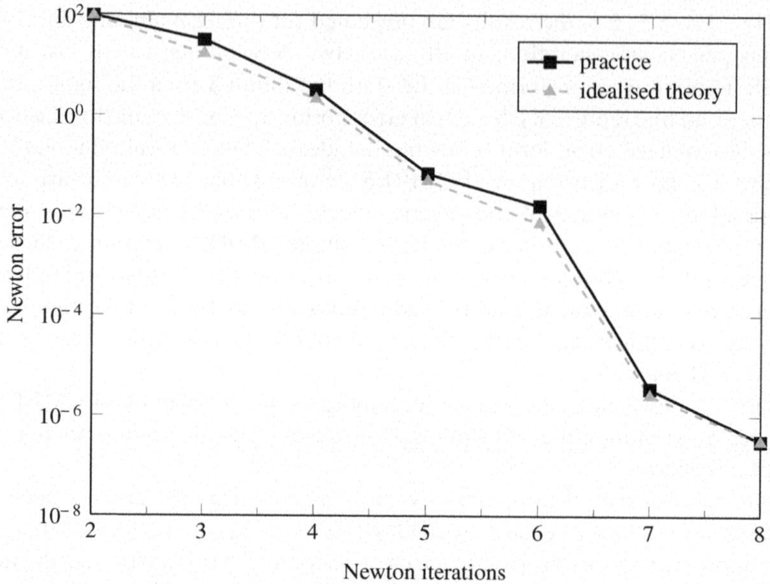

Fig. 5.5 GMRES iteration distribution 1, 1, 3, 4, 6, 3, 11, 3

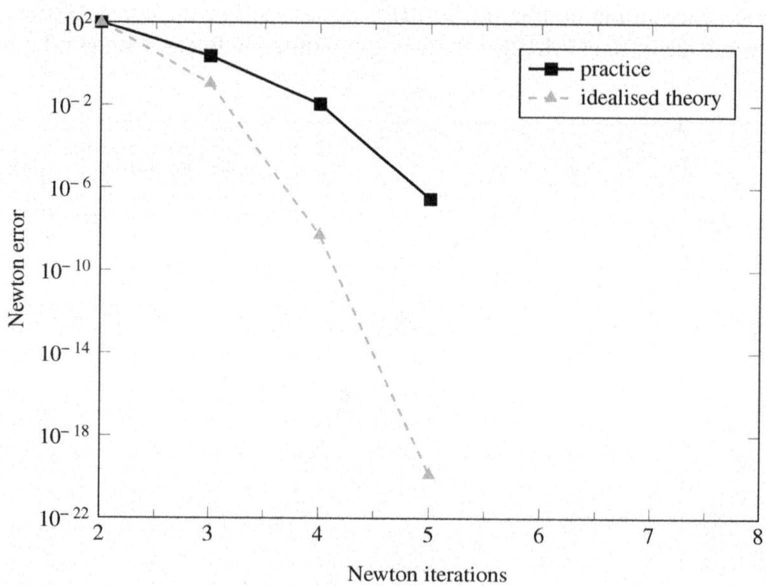

Fig. 5.6 GMRES iteration distribution 1, 1, 9, 19, 30

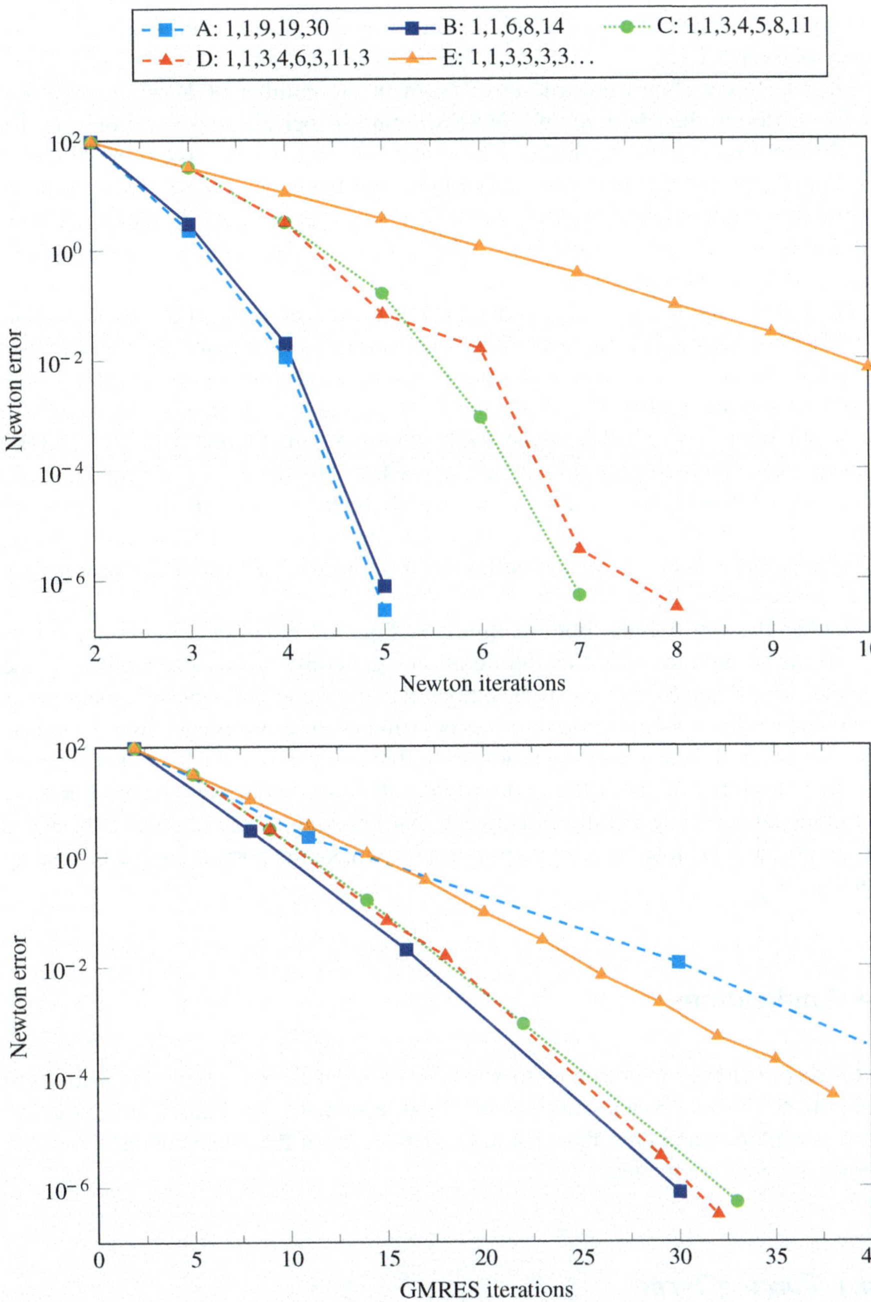

Fig. 5.7 Convergence in Newton and GMRES iterations

convergence in each Newton iteration. Thus Fig. 5.7 can be used to illustrate the theory of Sect. 5.2.1.

The top figure shows the true error norm in the number of Newton iterations, for five different distributions of GMRES iterations per Newton iteration, i.e., for five different sets of forcing terms. The graphs are as expected; the more GMRES iteration are performed per Newton iteration, the better the convergence. A naive interpretation might conclude that option (A) is the best of the considered choices, and that option (E) is by far the worst. However, this is too simple a conclusion, as illustrated by the bottom figure.

The bottom figure shows the convergence of the true error in the total number of GMRES iterations for the same five distributions. In this figure, the convergence of option (A) is worse than that of option (E), revealing that option (A) leads to a lot of oversolving. Option (E) is still the worst of the options that do not oversolve much, but not nearly as bad as suggested by the top figure. Options (B), (C), and (D) show approximately linear convergence, as predicted by the theory of Sect. 5.2.1. As the practical GMRES convergence is not exactly linear, nor exactly the same in each Newton iteration, the convergence of these options is not identical, and option (E) is still quite a bit worse. The strong influence of the near linear GMRES convergence is nonetheless very clear.

Neither the top nor the bottom figure in Fig. 5.7 tells the entire story on its own. If the set-up time of a Newton iteration—generally mostly determined by the calculation of J and $\mathbf{F}$—is very high compared to the computational cost of iterations of the linear solver, then the top figure approximates the convergence in the solution time. However, if these set-up costs are negligible compared to the linear solves, then it is the bottom figure that better approximates the convergence in the solution time. The practical truth is generally in between, but knowing which of these extremes a certain problem is closer to, can be important to make the correct choice of forcing terms.

5.4 Applications

In this section, ideas are presented to use the knowledge from the previous sections to design better inexact Newton algorithms. First, optimising the choice of the forcing terms is explored, and after that, possible adaptations of the linear solver within the Newton process are treated.

5.4.1 Forcing Terms

The ideas for choosing good forcing terms η_i rely on the expectation that, in Newton iteration i, both the unknown true error and its known measure $\|\mathbf{F}(\mathbf{x}_i)\|$ should reduce with an approximate factor η_i, as indicated by Theorem 5.1.

Theoretically, this knowledge can be used to choose the forcing terms adaptively by calculating $\|\mathbf{F}(\mathbf{x}_i + \mathbf{s}_i^k)\|$ in every linear iteration k, and checking whether the reduction in the norm of $\mathbf{F}$ is close enough to the reduction in the linear residual. Once the reduction in the norm of $\mathbf{F}$ starts lagging that of the linear residual, the linear solver is oversolving, and the next Newton iteration should be started. Obviously, this adaptive method only makes sense if $\|\mathbf{F}(\mathbf{x}_i + \mathbf{s}_i^k)\|$ can be evaluated cheaply compared to the cost of doing extra linear iterations, which is often not the case.

Theorem 5.1 can also be used to set a lower bound for the forcing terms. Assume that the aim is to solve up to $\|\mathbf{F}\| \leq \tau$. A forcing term $\eta_i = \frac{\tau}{\|\mathbf{F}(\mathbf{x}_i)\|}$ should be sufficient to reach that target, provided that there is no oversolving. Choosing η_i significantly smaller than that always leads to a waste of computational effort. Therefore, it makes sense to enforce, for some sensible choice of $\sigma \in (0, 1]$:

$$\eta_i \geq \sigma \frac{\tau}{\|\mathbf{F}(\mathbf{x}_i)\|}. \tag{5.40}$$

Knowledge of the computational cost to set up a Newton iteration, and of the convergence behaviour of the used iterative linear solver, can further help to choose better forcing terms. If the set-up cost of a Newton iteration is high, it then makes sense to choose smaller forcing terms to get the most out of each Newton iteration. Similarly, if the linear solver converges superlinearly, slightly smaller forcing terms may be preferred to maximise the benefit of this superlinear convergence. On the other hand if the set-up cost of a Newton iteration is low, then it may yield better results to keep the forcing terms a bit larger to prevent oversolving, especially if the linear solver does not converge superlinearly.

5.4.2 Linear Solver

Given a forcing term η_i, the choice of linear solver may be adapted to the value of this forcing term. For example, if it is expected that only a few linear iterations are needed, then GMRES is often the best choice due to its minimal residual property. On the other hand, if many linear iterations are anticipated it might be better to use Bi-CGSTAB or IDR(s). If the nonlinear problem is not too large, it may even be best to switch to a direct solver in later iterations, when η_i becomes very small.

Instead of changing the entire linear solver between Newton iterations, it is also an option to change just the preconditioning. For example, a higher quality type of preconditioner may be advantageous in later iterations. Alternatively, the preconditioner can be kept the same through multiple Newton iterations, depending on η_i.

References

1. Dembo, R.S., Eisenstat, S.C., Steihaug, T.: Inexact Newton methods. SIAM J. Numer. Anal. **19**(2), 400–408 (1982)
2. Ortega, J.M., Rheinboldt, W.C.: Iterative Solution of Nonlinear Equations in Several Variables. Academic Press, New York (1970)
3. Dennis Jr, J.E., Schnabel, R.B.: Numerical Methods for Unconstrained Optimization and Nonlinear Equations. Prentice Hall, New Jersey (1983)

Part II
Power System Analysis

Chapter 6
Power System Analysis

A power system is a system that provides for the generation, transmission, and distribution of electrical energy. Power systems are considered to be the largest and most complex man-made systems. As electrical energy is vital to our society, power systems have to satisfy the highest security and reliability standards. At the same time, minimising cost and environmental impact are important issues. Figure 6.1 shows a schematic representation of a power system.

Thermal power plants generate electrical power using heat, mostly from the combustion of fossil fuels, or from a nuclear reaction in the case of nuclear power plants. Most thermal power stations heat water to produce steam, which is then used to power turbines. Kinetic energy from these rotating devices is converted into electrical power by means of electromagnetic induction. Hydroelectric power plants run water through water turbines (typically located in dams), wind farms use wind turbines, and photovoltaic plants use solar panels to generate electrical power. Hydroelectric, wind, and solar power are examples of renewable energy, as they are generated from naturally replenished resources.

The transmission network connects the generating plants to substations near the consumers. It also performs the function of connecting different power pools, to reduce cost and increase reliability. High voltage alternating current (AC) is used to reduce voltage drops and power losses, and to increase capacity of the transmission lines. A three-phase system is used to reduce conductor material.

The distribution network connects the transmission network to the consumers. The distribution network operates at lower voltages than the transmission network, supplying three-phase AC to industrial consumers, and single-phase AC for common household consumption.

Power systems have to operate very close to a fixed frequency (50 Hz in most of the world, but typically 60 Hz in the Americas). Whenever an electrical appliance is turned on, the load on the power system increases. In the case of a thermal power plant, the extra power is taken from the kinetic energy of a rotating device, slowing down the rotation. Extra steam has to be fed to the turbines to keep the rotation at the desired frequency for the power system. Automated controls make it possible

R. Idema and D. J. P. Lahaye, *Computational Methods in Power System Analysis*, Atlantis Studies in Scientific Computing in Electromagnetics,
DOI: 10.2991/978-94-6239-064-5_6,

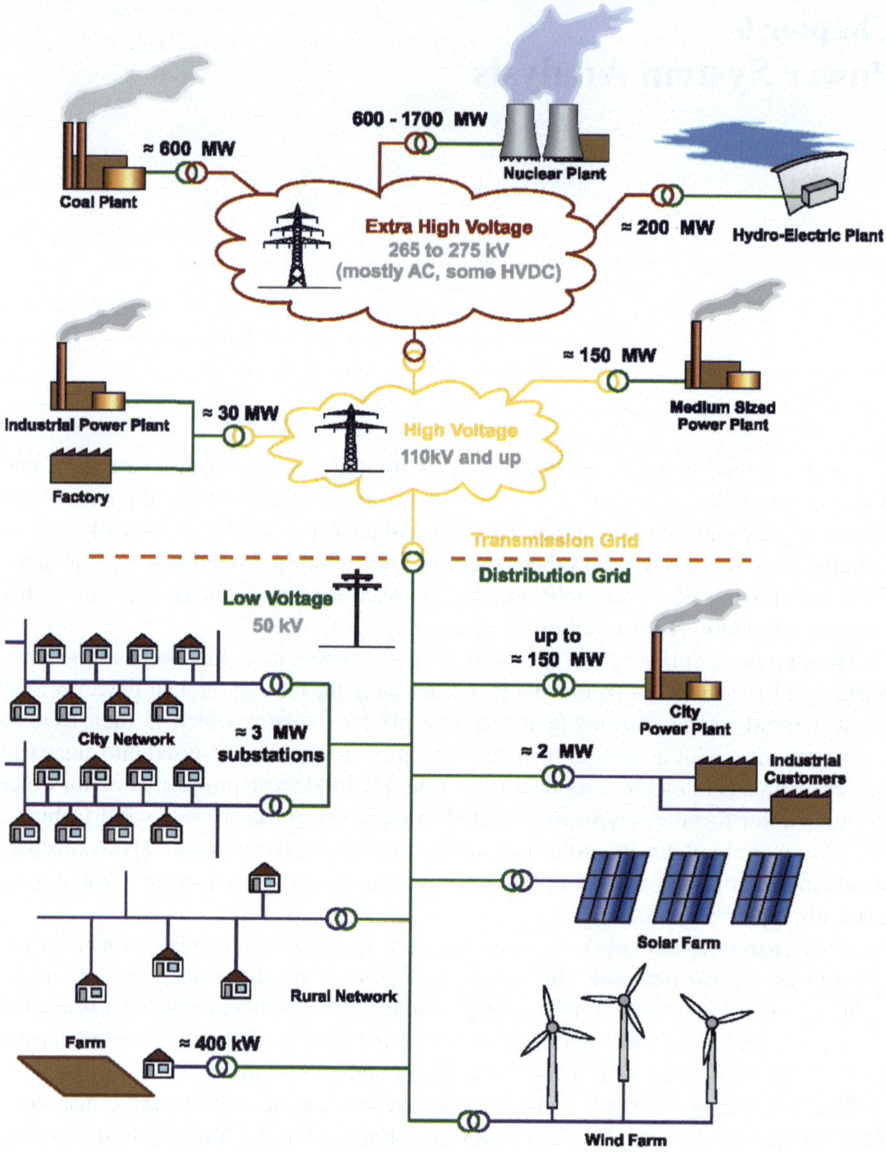

Fig. 6.1 Schematic representation of a power system [1]

for the power system to operate at near fixed frequency, making steady state power system models—where the frequency is regarded constant—a useful approximation of reality.

Steady state power system analysis, by means of simulations on mathematical models, plays an important role in both operational control and planning. This chapter first treats the required mathematical models of electrical power and power system components. Using these models, power flow and contingency analysis are treated. In power flow studies, the bus voltages in the power system are calculated, given the generation and consumption. Contingency analysis simulates equipment outages to determine if the system can still function reliably if such a contingency were to occur.

6.1 Electrical Power

To model a power system, models of the underlying quantities are needed, as well as mathematical relations between these quantities. This sections treats steady state models for the voltage, current, and power quantities, as well as quantities related to electrical resistance. Using these quantities, Ohm's law, and Kirchhoff's laws for AC circuits are treated.

6.1.1 Voltage and Current

In a power system in steady state, the voltage and current quantities can be assumed to be sinusoidal functions of time with constant frequency ω. By convention, these quantities are described using the cosine function, i.e.,

$$v(t) = V_{\max} \cos(\omega t + \delta_V) = \mathrm{Re}(V_{\max} e^{\iota\delta_V} e^{\iota\omega t}), \tag{6.1}$$

$$i(t) = I_{\max} \cos(\omega t + \delta_I) = \mathrm{Re}(I_{\max} e^{\iota\delta_I} e^{\iota\omega t}), \tag{6.2}$$

where ι is the imaginary unit, and Re is the operator that takes the real part.

Since the frequency ω is assumed constant in a steady state power system, the term $e^{\iota\omega t}$ is not needed to describe the voltage or current. The remaining complex quantities $V = V_{\max} e^{\iota\delta_V}$ and $I = I_{\max} e^{\iota\delta_I}$ are independent of the time t, and are called the phasor representation of the voltage and current respectively. These quantities are used to represent the voltage and current in circuit theory. In power system theory, instead the effective phasor representation is used:

$$V = |V|\, e^{\iota\delta_V}, \text{ with } |V| = \frac{V_{\max}}{\sqrt{2}}, \tag{6.3}$$

$$I = |I|\, e^{\iota\delta_I}, \text{ with } |I| = \frac{I_{\max}}{\sqrt{2}}. \tag{6.4}$$

Note that $|V|$ and $|I|$ are the RMS values of $v(t)$ and $i(t)$, and that the effective phasors differ from the circuit theory phasors by a factor $\sqrt{2}$.

As this book is about steady state power system calculations, V and I are used to denote the effective voltage and current phasors, as defined above.

6.1.2 Complex Power

Using the voltage and current Eqs. (6.1) and (6.2), the reference time can be chosen such that the voltage can be written as $v(t) = V_{\max}\cos(\omega t)$, and the current as $i(t) = I_{\max}\cos(\omega t - \phi)$. The quantity $\phi = \delta_V - \delta_I$ is called the power factor angle, and $\cos\phi$ the power factor.

The instantaneous power $p(t)$ is then given by

$$\begin{aligned} p(t) &= v(t)i(t) \\ &= \sqrt{2}\,|V|\cos(\omega t)\sqrt{2}\,|I|\cos(\omega t - \phi) \\ &= 2\,|V|\,|I|\cos(\omega t)\cos(\omega t - \phi) \\ &= 2\,|V|\,|I|\cos(\omega t)\left[\cos\phi\cos(\omega t) + \sin\phi\sin(\omega t)\right] \\ &= |V|\,|I|\left[2\cos\phi\cos^2(\omega t) + 2\sin\phi\sin(\omega t)\cos(\omega t)\right] \\ &= |V|\,|I|\cos\phi\left[2\cos^2(\omega t)\right] + |V|\,|I|\sin\phi\left[2\sin(\omega t)\cos(\omega t)\right] \\ &= |V|\,|I|\cos\phi\left[1 + \cos(2\omega t)\right] + |V|\,|I|\sin\phi\left[\sin(2\omega t)\right] \\ &= P\left[1 + \cos(2\omega t)\right] + Q\left[\sin(2\omega t)\right], \end{aligned} \tag{6.5}$$

where $P = |V|\,|I|\cos\phi$, and $Q = |V|\,|I|\sin\phi$.

Thus the instantaneous power is the sum of a unidirectional component that is sinusoidal with average value P and amplitude P, and a component of alternating direction that is sinusoidal with average 0 and amplitude Q. Note that integrating the instantaneous power over a time period $T = \frac{2\pi}{\omega}$ gives

$$\frac{1}{T}\int_0^T p(t)dt = P. \tag{6.6}$$

The magnitude P is called the active power, or real power, or average power, and is measured in W (watts). The magnitude Q is called the reactive power, or imaginary power, and is measured in var (volt-ampere reactive).

Using the complex representation of voltage and current, we can write

$$P = |V|\,|I|\cos\phi = \mathrm{Re}(|V|\,|I|\,e^{\imath(\delta_V - \delta_I)}) = \mathrm{Re}(V\overline{I}), \tag{6.7}$$

$$Q = |V|\,|I|\sin\phi = \mathrm{Im}(|V|\,|I|\,e^{\imath(\delta_V - \delta_I)}) = \mathrm{Im}(V\overline{I}), \tag{6.8}$$

where $\overline{I}$ is the complex conjugate of I. Thus we can define the complex power in AC circuits as

$$S = P + \iota Q = V\bar{I}, \tag{6.9}$$

where S is measured in VA (volt-ampere).

Note that strictly speaking VA and var are the same unit as W, however it is useful to use the different unit names to distinguish between the measured quantities.

6.1.3 Impedance and Admittance

Impedance is the extension of the notion of resistance to AC circuits. It is a measure of opposition to a sinusoidal current. The impedance is denoted by

$$Z = R + \iota X, \tag{6.10}$$

and measured in ohms (Ω). The real part $\operatorname{Re} Z = R \geq 0$ is called the resistance, and the imaginary part $\operatorname{Im} Z = X$ the reactance. If $X > 0$ the reactance is called inductive and we can write $\iota X = \iota \omega L$, where $L > 0$ is the inductance. If $X < 0$ the reactance is called capacitive and we write $\iota X = \frac{1}{\iota \omega C}$, where $C > 0$ is the capacitance.

The admittance

$$Y = G + \iota B \tag{6.11}$$

is the inverse of the impedance and is measured in siemens (S), i.e.,

$$Y = \frac{1}{Z} = \frac{R}{|Z|^2} - \iota \frac{X}{|Z|^2}. \tag{6.12}$$

The real part $G = \operatorname{Re} Y = \frac{R}{|Z|^2} \geq 0$ is called the conductance, while the imaginary part $B = \operatorname{Im} Y = -\frac{X}{|Z|^2}$ is called the susceptance.

The voltage drop over an impedance Z is equal to $V = ZI$. This is the extension of Ohm's law to AC circuits. Alternatively, using the admittance, we can write

$$I = YV. \tag{6.13}$$

Using Ohm's law, we find that the power consumed by an impedance Z is

$$S = V\bar{I} = ZI\bar{I} = |I|^2 Z = |I|^2 R + \iota |I|^2 X. \tag{6.14}$$

6.1.4 *Kirchhoff's Circuit Laws*

Kirchhoff's circuit laws are used to calculate the voltage and current in electrical circuits.

Kirchhoff's current law (KCL): At any point in the circuit, the sum of currents flowing towards that point is equal to the sum of currents flowing away from that point, i.e., $\sum_k I_k = 0$.

Kirchhoff's voltage law (KVL): The directed sum of the electrical potential differences around any closed circuit is zero, i.e., $\sum_k V_k = 0$.

6.2 Power System Model

Power systems are modelled as a network of buses (nodes) and branches (edges). At each bus i, four electrical quantities are of importance:

$$\begin{aligned} |V_i| &: \text{voltage magnitude}, \\ \delta_i &: \text{voltage phase angle}, \\ P_i &: \text{injected active power}, \\ Q_i &: \text{injected reactive power}. \end{aligned}$$

Each bus can hold a number of electrical devices. The bus is named according to the electrical magnitudes specified at that bus, see Table 6.1.

Local distribution networks are usually connected to the transmission network at a single bus. In steady state power system models, such networks generally get aggregated in that connecting bus, which then gets assigned the total load of the distribution network.

Further, balanced three-phase systems are represented by one-line diagrams of equivalent single-phase systems, and voltage and current quantities are represented in per unit (p.u.). For more details see for example [2, 3].

Table 6.1 Bus types with electrical magnitudes

Bus type	Specified	Unknown
Load bus or PQ bus	P_i, Q_i	$\lvert V_i\rvert, \delta_i$
Generator bus or PV bus	$P_i, \lvert V_i\rvert$	Q_i, δ_i
Slack bus or swing bus	$\delta_i, \lvert V_i\rvert$	P_i, Q_i

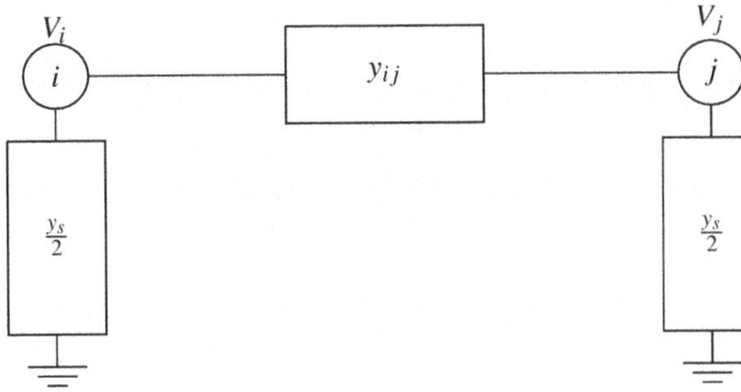

Fig. 6.2 Transmission line model

6.2.1 Generators, Loads, and Transmission Lines

A physical generator usually has P and $|V|$ controls and thus specifies these magnitudes. Likewise, a load will have a negative injected active power P specified, as well as a reactive power Q. However, the name of the bus does not necessarily indicate what type of devices it consists of. A wind turbine, for example, is a generator but does not have PV controls. Instead, it is modelled as a load bus with positive injected active power P. When a PV generator and a PQ load are connected to the same bus, the result is a PV bus with a voltage amplitude equal to that of the generator, and an active power equal to the sum of the active power of the generator and the load. Also, there may be buses without a generator or load connected, such as transmission substations, which are modelled as loads with $P = Q = 0$.

In any practical power system there are system losses. These losses have to be taken into account, but since they depend on the power flow they are not known in advance. A generator bus has to be assigned that will compensate for the difference between the total specified generation and the total specified load plus losses. This bus is called the slack bus, or swing bus. Obviously it is not possible to specify the active power P for this bus. Instead the voltage magnitude $|V|$ and angle δ are specified. Note that δ is merely the reference phase to which the other phase angles are measured. As such, it is common to set $\delta = 0$ for the slack bus.

Transmission lines (and cables) are represented by branches that connect the buses in the power system. From a modelling viewpoint, branches define how to relate buses with Kirchhoff's circuit laws. Transmission lines generally incur losses on the transported power and must be modelled as such.

A transmission line from bus i to bus j has some impedance. This impedance is modelled as a single total impedance quantity z_{ij} on the branch. The admittance then is $y_{ij} = \frac{1}{z_{ij}}$. Further, there is a shunt admittance from the line to the neutral ground. This shunt admittance is modelled as a total shunt admittance quantity y_s that is split evenly between bus i and bus j. Figure 6.2 shows a schematic representation of the transmission line (or cable) model.

It is usually assumed that there is no conductance from the line to the ground. This means that the shunt admittance is due only to the electrical field between line

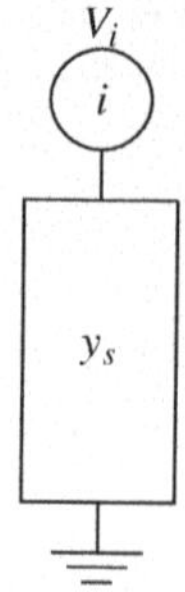

Fig. 6.3 Shunt model

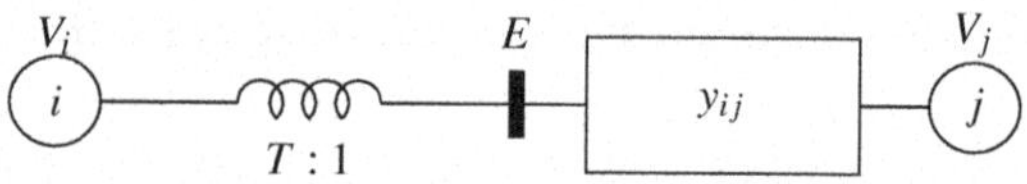

Fig. 6.4 Transformer model

and ground, and is thus a capacitive susceptance, i.e., $y_s = \iota b_s$, with $b_s \geq 0$. For this reason, the shunt admittance y_s is also sometimes referred to as the shunt susceptance b_s. See also the notes about modelling shunts in Sect. 6.2.2.

6.2.2 Shunts and Transformers

Two other devices commonly found in power systems are shunts and transformers. Shunt capacitors can be used to inject reactive power, resulting in a higher node voltage, while shunt inductors consume reactive power, lowering the node voltage. Transformers are used to step-up the voltage to a higher level, or step-down to a lower level. A phase shifting transformer (PST) can also change the voltage phase angle. Some transformers have different taps, and can use tap changing to adjust the transformer ratio.

A shunt is modelled as a reactance $z_s = \iota x_s$ between the bus and the ground, see Fig. 6.3. The shunt admittance thus is $y_s = \frac{1}{z_s} = -\iota \frac{1}{x_s} = \iota b_s$. If $x_s > 0$ the shunt is inductive, if $x_s < 0$ the shunt is capacitive. Note that the shunt susceptance b_s has the opposite sign of the shunt reactance x_s.

Transformers can be modelled as depicted in Fig. 6.4, where $T : 1$ is the transformer ratio. The modulus of T determines the change in voltage magnitude. This value is usually around 1, because the better part of the differences in voltage levels are incorporated through the per unit system. The argument of T determines the shift of the voltage phase angle.

6.2.3 Admittance Matrix

The admittance matrix Y is a matrix that relates the injected current at each bus to bus voltages, such that

$$\mathbf{I} = Y\mathbf{V}, \tag{6.15}$$

where $\mathbf{I}$ is the vector of injected currents at each bus, and $\mathbf{V}$ is the vector of bus voltages. This is in fact Ohm's law (6.13) in matrix form. As such we can define the impedance matrix as $Z = Y^{-1}$.

To calculate the admittance matrix Y, we look at the injected current I_i at each bus i. Let I_{ij} denote the current flowing from bus i in the direction of bus $j \neq i$, or to the ground in case of a shunt. Applying Kirchhoff's current law now gives

$$I_i = \sum_k I_{ik}. \tag{6.16}$$

Let y_{ij} denote the admittance of the branch between bus i and j, with $y_{ij} = 0$ if there is no branch between these buses. For a simple transmission line from bus i to bus j, i.e., without shunt admittance, Ohm's law states that

$$I_{ij} = y_{ij}\left(V_i - V_j\right), \text{ and } I_{ji} = -I_{ij}, \tag{6.17}$$

or in matrix notation:

$$\begin{bmatrix} I_{ij} \\ I_{ji} \end{bmatrix} = y_{ij} \begin{bmatrix} 1 & -1 \\ -1 & 1 \end{bmatrix} \begin{bmatrix} V_i \\ V_j \end{bmatrix}. \tag{6.18}$$

Now suppose that there is a shunt s connected to bus i. Then, according to Eq. (6.16), an extra term I_{is} is added to the injected current I_i. From Fig. 6.3, it is clear that

$$I_{is} = y_s\left(V_i - 0\right) = y_s V_i. \tag{6.19}$$

This means that in the admittance matrix an extra term y_s has to be added to Y_{ii}. Recall that $y_s = \imath b_s$, and that the sign of b_s depends on the shunt being inductive or capacitive.

Knowing how to deal with shunts, it is now easy to incorporate the line shunt model as depicted in Fig. 6.2. For a transmission line between the buses i and j, half of the line shunt admittance of that line, i.e., $\frac{y_s}{2}$, has to be added to both Y_{ii} and Y_{jj} in the admittance matrix. For a transmission line with shunt admittance y_s, we thus find

$$\begin{bmatrix} I_{ij} \\ I_{ji} \end{bmatrix} = \left(y_{ij} \begin{bmatrix} 1 & -1 \\ -1 & 1 \end{bmatrix} + y_s \begin{bmatrix} \frac{1}{2} & 0 \\ 0 & \frac{1}{2} \end{bmatrix} \right) \begin{bmatrix} V_i \\ V_j \end{bmatrix}. \tag{6.20}$$

The impact of a transformer between the buses i and j on the admittance matrix, can be derived from the model depicted in Fig. 6.4.

Let E be the voltage induced by the transformer, then

$$V_i = TE. \tag{6.21}$$

The current flowing from bus j in the direction of the transformer device (towards bus i) then is

$$I_{ji} = y_{ij}\left(V_j - E\right) = y_{ij}\left(V_j - \frac{V_i}{T}\right). \tag{6.22}$$

Conservation of power within the transformer gives

$$V_i \overline{I}_{ij} = -E\overline{I}_{ji} \Leftrightarrow T\overline{I}_{ij} = -\overline{I}_{ji} \Leftrightarrow \overline{T} I_{ij} = -I_{ji}. \tag{6.23}$$

Therefore, the current flowing from bus i in the direction of the transformer device (towards bus j) is

$$I_{ij} = -\frac{I_{ji}}{\overline{T}} = y_{ij}\left(\frac{V_i}{|T|^2} - \frac{V_j}{\overline{T}}\right). \tag{6.24}$$

The total contribution of a branch between bus i and bus j to the admittance matrix, thus becomes

$$\begin{bmatrix} I_{ij} \\ I_{ji} \end{bmatrix} = \left(y_{ij} \begin{bmatrix} \frac{1}{|T|^2} & -\frac{1}{\overline{T}} \\ -\frac{1}{T} & 1 \end{bmatrix} + y_s \begin{bmatrix} \frac{1}{2} & 0 \\ 0 & \frac{1}{2} \end{bmatrix} \right) \begin{bmatrix} V_i \\ V_j \end{bmatrix}, \tag{6.25}$$

where $T = 1$ if the branch is not a transformer.

The admittance matrix Y can now be constructed as follows. Start with a diagonal matrix with the shunt admittance value on diagonal element i for each bus i that has a shunt device, and 0 on each diagonal element for which the corresponding bus has no shunt device. Then, for each branch add its contribution to the matrix according to Eq. (6.25).

6.3 Power Flow

The power flow problem, or load flow problem, is the problem of computing the flow of electrical power in a power system in steady state. In practice, this amounts to calculating the voltage in each bus of the power system. Once the bus voltages are known, the other electrical quantities are easy to compute. The power flow problem has many applications in power system operation and planning, and is treated in many books on power systems, see for example [2–4].

Mathematical equations for the power flow problem can be obtained by combining the complex power equation (6.9), with Ohm's law (6.15). This gives

$$S_i = V_i \overline{I}_i = V_i \left(\overline{Y\mathbf{V}}\right)_i = V_i \sum_{k=1}^{N} \overline{Y}_{ik} \overline{V}_k, \tag{6.26}$$

where S_i is the injected power at bus i, I_i the current through bus i, V_i the bus voltage, Y is the admittance matrix, and N is the number of buses in the power system.

The admittance matrix Y is easy to obtain, and generally very sparse. Therefore, a formulation using the admittance matrix has preference over formulations using the impedance matrix Z, which is generally a lot harder to obtain and not sparse.

In Chap. 7 two traditional methods to solve the power flow problem (6.26) are treated. In Chap. 8 we investigate power flow solvers based on Newton-Krylov methods, and show that such solvers scale much better in the problem size, making them much faster than the traditional methods for large power flow problems.

6.4 Contingency Analysis

Contingency analysis is the act of identifying changes in a power system that have some non-negligible chance of unplanned occurrence, and analysing the impact of these contingencies on power system operation. The contengencies most commonly studied are single generator and branch outages.

A power system that still operates properly on the occurrence of any single contingency, is called $n-1$ secure. In some cases $n-2$ security analysis is desired, i.e., analysis of the impact of any two contingencies happening simultaneously.

Contingency analysis generally involves solving many power flow problems, in which the contingencies have been modelled. In Chap. 9 we investigate how Newton-Krylov power flow solvers can be exploited to speed up contingency analysis calculations.

References

1. Wikipedia: Electricity Grid Schematic English—Wikipedia, the free encyclopedia (2010). http://en.wikipedia.org/wiki/File:Electricity_Grid_Schematic_English.svg
2. Bergen, A.R., Vittal, V.: Power Systems Analysis, 2nd edn. Prentice Hall, New Jersey (2000)
3. Schavemaker, P., van der Sluis, L.: Electrical Power System Essentials. Wiley, Chichester (2008)
4. Powell, L.: Power System Load Flow Analysis. McGraw-Hill, USA (2004)

Chapter 7
Traditional Power Flow Solvers

As long as there have been power systems, there have been power flow studies. This chapter discusses the two traditional methods to solve power flow problems: Newton power flow and Fast Decoupled Load Flow (FDLF).

Newton power flow is described in Sect. 7.1. The concept of the power mismatch function is treated, and the corresponding Jacobian matrix is derived. Further, it is detailed how to treat different bus types within the Newton power flow method.

Fast Decoupled Load Flow is treated in Sect. 7.2. The FDLF method can be seen as a clever approximation of Newton power flow. Instead of the Jacobian matrix, an approximation based on the practical properties of power flow problems is calculated once, and used throughout all iterations.

Finally, Sect. 7.3 discusses convergence and computational properties of the two methods, and Sect. 7.4 describes how Newton power flow and FDLF can be interpreted as basic Newton-Krylov methods, motivating how Newton-Krylov methods can be used to improve on these traditional power flow solvers.

7.1 Newton Power Flow

Newton power flow uses the Newton-Raphson method to solve the power flow problem. Traditionally, a direct solver is used to solve the linear system of equations (4.6) that arises in each iteration of the Newton method [1, 2].

In order to use the Newton-Raphson method, the power flow equations have to be written in the form $\mathbf{F}(\mathbf{x}) = \mathbf{0}$. The common procedure to get such a form is described in Sect. 7.1.1. This procedure leads to a function $\mathbf{F}(\mathbf{x})$ called the power mismatch function. The power mismatch function contains the the injected active power P_i and reactive power Q_i at each bus, while the vector parameter $\mathbf{x}$ consists of the voltage angles δ_i and voltage magnitudes $|V_i|$. Newton power flow usually uses a flat start, meaning that the initial iterate has $\delta_i = 0$ and $|V_i| = 1$ for all i.

R. Idema and D. J. P. Lahaye, *Computational Methods in Power System Analysis*,
Atlantis Studies in Scientific Computing in Electromagnetics,
DOI: 10.2991/978-94-6239-064-5_7, © Atlantis Press and the authors 2014

Another element required for the Newton-Raphson method, is the Jacobian matrix $J(\mathbf{x})$. In Sect. 7.1.2 the Jacobian matrix of the power mismatch function is derived. Further, it is shown that this matrix can be computed cheaply from the building blocks used in the evaluation of the power mismatch function.

For load buses the voltage angle δ_i and voltage magnitude $|V_i|$ are the unknowns, see Table 6.1 (p. 52). However, for generator buses the voltage magnitude δ_i is known, while the injected reactive power Q_i is unknown. And for the slack bus, the entire voltage phasor is known, while the injected power is unknown. Thus, the power mismatch function $\mathbf{F}(\mathbf{x})$ is not simply a known function in an unknown parameter. Section 7.1.3 deals with the steps needed for each of the different bus types, to be able to apply the Newton-Raphson method to the power mismatch function.

7.1.1 Power Mismatch Function

Recall from Sect. 6.3 that the power flow problem can be described by the power flow equations

$$S_i = V_i \sum_{k=1}^{N} \overline{Y}_{ik} \overline{V}_k. \tag{7.1}$$

As it is not possible to treat the voltage phasors V_i as variables of the problem for the slack bus and generator buses, it makes sense to rewrite the N complex nonlinear equations of Eq. (6.26) as $2N$ real nonlinear equations in the real quantities P_i, Q_i, $|V_i|$, and δ_i.

Substituting $V_i = |V_i| e^{\iota\delta_i}$, $Y = G + \iota B$, and $\delta_{ij} = \delta_i - \delta_j$ into the power flow equations (7.1) gives

$$\begin{aligned} S_i &= |V_i| e^{\iota\delta_i} \sum_{k=1}^{N} (G_{ik} - \iota B_{ik}) |V_k| e^{-\iota\delta_k} \\ &= \sum_{k=1}^{N} |V_i| |V_k| (\cos\delta_{ik} + \iota \sin\delta_{ik}) (G_{ik} - \iota B_{ik}) . \end{aligned} \tag{7.2}$$

Now define the real vector $\mathbf{x}$ of voltage variables as

$$\mathbf{x} = [\delta_1, \ldots, \delta_N, |V_1|, \ldots, |V_N|]^T . \tag{7.3}$$

For the purpose of notational comfort, further define the matrix functions $\mathcal{P}(\mathbf{x})$ and $\mathcal{Q}(\mathbf{x})$ with entries

$$\mathcal{P}_{ij}(\mathbf{x}) = |V_i| \left|V_j\right| \left(G_{ij} \cos\delta_{ij} + B_{ij} \sin\delta_{ij}\right), \tag{7.4}$$

$$\mathcal{Q}_{ij}(\mathbf{x}) = |V_i| \left|V_j\right| \left(G_{ij} \sin\delta_{ij} - B_{ij} \cos\delta_{ij}\right), \tag{7.5}$$

and the vector functions $\mathbf{P}(\mathbf{x})$ and $\mathbf{Q}(\mathbf{x})$ with entries

$$P_i(\mathbf{x}) = \sum_k \mathcal{P}_{ik}(\mathbf{x}), \tag{7.6}$$
$$Q_i(\mathbf{x}) = \sum_k \mathcal{Q}_{ik}(\mathbf{x}). \tag{7.7}$$

Note that $\mathcal{P}_{ij}(\mathbf{x}) = \mathcal{Q}_{ij}(\mathbf{x}) = 0$ for each pair of buses $i \neq j$ that is not connected by a branch.

Using the above definitions, Eq. (7.2) can be written as

$$\mathbf{S} = \mathbf{P}(\mathbf{x}) + \iota\mathbf{Q}(\mathbf{x}). \tag{7.8}$$

Now, the power mismatch function $\mathbf{F}$ is the real vector function

$$\mathbf{F}(\mathbf{x}) = \begin{bmatrix} \mathbf{P} - \mathbf{P}(\mathbf{x}) \\ \mathbf{Q} - \mathbf{Q}(\mathbf{x}) \end{bmatrix}, \tag{7.9}$$

and the power flow problem can be written as the system of nonlinear equations

$$\mathbf{F}(\mathbf{x}) = \mathbf{0}. \tag{7.10}$$

7.1.2 Jacobian Matrix

The Jacobian matrix $J(\mathbf{x})$ of a function $\mathbf{F}(\mathbf{x})$, is the matrix of all first order partial derivatives of that function. The Jacobian matrix of the power mismatch function has the following structure, where $P_i(\mathbf{x})$ and $Q_i(\mathbf{x})$ are as in Eqs. (7.6) and (7.7) respectively:

$$J(\mathbf{x}) = -\left[\begin{array}{ccc|ccc} \frac{\partial P_1}{\partial \delta_1}(\mathbf{x}) & \cdots & \frac{\partial P_1}{\partial \delta_N}(\mathbf{x}) & \frac{\partial P_1}{\partial |V_1|}(\mathbf{x}) & \cdots & \frac{\partial P_1}{\partial |V_N|}(\mathbf{x}) \\ \vdots & \ddots & \vdots & \vdots & \ddots & \vdots \\ \frac{\partial P_N}{\partial \delta_1}(\mathbf{x}) & \cdots & \frac{\partial P_N}{\partial \delta_N}(\mathbf{x}) & \frac{\partial P_N}{\partial |V_1|}(\mathbf{x}) & \cdots & \frac{\partial P_N}{\partial |V_N|}(\mathbf{x}) \\ \hline \frac{\partial Q_1}{\partial \delta_1}(\mathbf{x}) & \cdots & \frac{\partial Q_1}{\partial \delta_N}(\mathbf{x}) & \frac{\partial Q_1}{\partial |V_1|}(\mathbf{x}) & \cdots & \frac{\partial Q_1}{\partial |V_N|}(\mathbf{x}) \\ \vdots & \ddots & \vdots & \vdots & \ddots & \vdots \\ \frac{\partial Q_N}{\partial \delta_1}(\mathbf{x}) & \cdots & \frac{\partial Q_N}{\partial \delta_N}(\mathbf{x}) & \frac{\partial Q_N}{\partial |V_1|}(\mathbf{x}) & \cdots & \frac{\partial Q_N}{\partial |V_N|}(\mathbf{x}) \end{array}\right]. \tag{7.11}$$

Note that the Jacobian matrix (7.11) consist of the negated first order derivatives of $P_i(\mathbf{x})$ and $Q_i(\mathbf{x})$, but that the Newton-Raphson method uses the negated Jacobian. Therefore, the coefficient matrix of the linear system solved in each Newton iteration, consists of the first order derivatives of $P_i(\mathbf{x})$ and $Q_i(\mathbf{x})$. These partial derivatives are derived below, where it is assumed that $i \neq j$ whenever applicable.

The first order partial derivatives of $P_i(\mathbf{x})$ and $Q_i(\mathbf{x})$ are:

$$\frac{\partial P_i}{\partial \delta_j}(\mathbf{x}) = |V_i|\,|V_j|\left(G_{ij}\sin\delta_{ij} - B_{ij}\cos\delta_{ij}\right) = \mathcal{Q}_{ij}(\mathbf{x}), \tag{7.12}$$

$$\begin{aligned}\frac{\partial P_i}{\partial \delta_i}(\mathbf{x}) &= \sum_{k\neq i} |V_i|\,|V_k|\,(-G_{ik}\sin\delta_{ik} + B_{ik}\cos\delta_{ik}) \\ &= -\sum_{k\neq i} \mathcal{Q}_{ik}(\mathbf{x}) = -Q_i(\mathbf{x}) - |V_i|^2 B_{ii},\end{aligned} \tag{7.13}$$

$$\frac{\partial Q_i}{\partial \delta_j}(\mathbf{x}) = |V_i|\,|V_j|\left(-G_{ij}\cos\delta_{ij} - B_{ij}\sin\delta_{ij}\right) = -\mathcal{P}_{ij}(\mathbf{x}), \tag{7.14}$$

$$\begin{aligned}\frac{\partial Q_i}{\partial \delta_i}(\mathbf{x}) &= \sum_{k\neq i} |V_i|\,|V_k|\,(G_{ik}\cos\delta_{ik} + B_{ik}\sin\delta_{ik}) \\ &= \sum_{k\neq i} \mathcal{P}_{ik}(\mathbf{x}) = P_i(\mathbf{x}) - |V_i|^2 G_{ii},\end{aligned} \tag{7.15}$$

$$\frac{\partial P_i}{\partial |V_j|}(\mathbf{x}) = |V_i|\left(G_{ij}\cos\delta_{ij} + B_{ij}\sin\delta_{ij}\right) = \frac{\mathcal{P}_{ij}(\mathbf{x})}{|V_j|}, \tag{7.16}$$

$$\begin{aligned}\frac{\partial P_i}{\partial |V_i|}(\mathbf{x}) &= 2\,|V_i|\,G_{ii} + \sum_{k\neq i} |V_k|\,(G_{ik}\cos\delta_{ik} + B_{ik}\sin\delta_{ik}) \\ &= 2\,|V_i|\,G_{ii} + \sum_{k\neq i} \frac{\mathcal{P}_{ik}(\mathbf{x})}{|V_i|} = \frac{P_i(\mathbf{x}) + |V_i|^2 G_{ii}}{|V_i|},\end{aligned} \tag{7.17}$$

$$\frac{\partial Q_i}{\partial |V_j|}(\mathbf{x}) = |V_i|\left(G_{ij}\sin\delta_{ij} - B_{ij}\cos\delta_{ij}\right) = \frac{\mathcal{Q}_{ij}(\mathbf{x})}{|V_j|}, \tag{7.18}$$

$$\begin{aligned}\frac{\partial Q_i}{\partial |V_i|}(\mathbf{x}) &= -2\,|V_i|\,B_{ii} + \sum_{k\neq i} |V_k|\,(G_{ik}\sin\delta_{ik} - B_{ik}\cos\delta_{ik}) \\ &= -2\,|V_i|\,B_{ii} + \sum_{k\neq i} \frac{\mathcal{Q}_{ik}(\mathbf{x})}{|V_i|} = \frac{Q_i(\mathbf{x}) - |V_i|^2 B_{ii}}{|V_i|}.\end{aligned} \tag{7.19}$$

Observe that the Jacobian matrix entries consist of the same building blocks $\mathcal{P}_{ij}$ and $\mathcal{Q}_{ij}$ as the power mismatch function $\mathbf{F}$. This means that whenever the power mismatch function is evaluated, the Jacobian matrix can be calculated at relatively little extra computational cost.

7.1.3 Handling Different Bus Types

Which of the values P_i, Q_i, $|V_i|$, and δ_i are specified, and which are not, depends on the associated buses, see Table 6.1 (p. 52).

Dealing with the fact that some elements in **P** and **Q** are not specified is easy. The equations corresponding to these unknowns can simply be dropped from the problem formulation. The unknown voltages in **x** can be calculated from the remaining equations, after which the unknown power values follow from evaluating the corresponding entries of $\mathbf{P}(\mathbf{x})$ and $\mathbf{Q}(\mathbf{x})$.

Dealing with specified voltage values is less straight-forward. Recall that the Newton-Raphson method is an iterative process that, in each iteration, calculates a vector $\mathbf{s}_i$ and sets the new iterate to be $\mathbf{x}_{i+1} = \mathbf{x}_i + \mathbf{s}_i$. Now, if some entries of **x** are known—as is the case for generator buses and the slack bus—then the best value for the corresponding entry of the update vector $\mathbf{s}_i$ is clearly 0.

To ensure that the update for known voltage values is indeed 0, these entries in the update vector, and the corresponding columns of the coefficient matrix, can simply be dropped. Thus for every generator bus, one unknown in the update vector and one column in the coefficient matrix are dropped, whereas for the slack bus two of each are dropped.

The amount of nonlinear equations dropped from the problem, is always equal to the amount of variables, and corresponding columns, that are dropped from the linear systems. Therefore, the linear systems that are actually solved have a square coefficient matrix of size $2N - N_G - 2 = 2N_L + N_G$, where N_L is the number of load buses, and N_G is the number of generator buses in the power system.

Another method to deal with different bus types is not to eliminate any rows or columns from the problem. Instead the linear systems are built normally, except for the linear equations that correspond to power values that are not specified. For these equations, the right-hand side value and all off-diagonal entries are set to 0, while the diagonal entry is set to 1. Or, the diagonal entry can be set to some very large number, in which case the off-diagonal entries can be kept as they would have been.

This method also ensures that the update for known voltage values is 0 in each iteration. The linear systems that have to be solved are of size $2N$, and thus larger than in the previous method. However, the structure of the matrix can be made independent of the bus types. This means that the matrix structure can be kept the same between runs that change the type of one or more buses. Bus-type switching is used for example to ensure that reactive power limits of generators are satisfied.

7.2 Fast Decoupled Load Flow

Fast Decoupled Load Flow (FDLF) is an approximation of Newton power flow, based on practical properties of power flow problems. The general FDLF method is shown in Algorithm 7.1.

The original derivation of the FDLF method is presented in Sect. 7.2.1, and in Sect. 7.2.2 notes on dealing with shunts and transformers are added. Finally, Sect. 7.2.3 treats different choices for the matrices B' and B'', called schemes, and explains how the BX and XB schemes can be interpreted as an approximation

Algorithm 7.1 Fast Decoupled Load Flow

```
calculate the matrices B' and B''
calculate LU factorisation of B' and B''
given initial iterates δ and |V|
while not converged do
    solve B'Δδ = ΔP(δ, |V|)
    update δ := δ + Δδ
    solve B''Δ|V| = ΔQ(δ, |V|)
    update |V| := |V| + Δ|V|
end while
```

of Newton power flow using the Schur complement. The techniques described in Sect. 7.1.3 can again be used to handle the different bus types.

7.2.1 Classical Derivation

In Fast Decoupled Load Flow, the assumption is made that for all i, j

$$\delta_{ij} \approx 0, \tag{7.20}$$
$$|V_i| \approx 1. \tag{7.21}$$

In the original derivation in [3], it is further assumed that

$$\left|G_{ij}\right| \ll \left|B_{ij}\right|. \tag{7.22}$$

Using assumption (7.20), the following approximations can be made:

$$\mathcal{P}_{ij}(\mathbf{x}) = |V_i| \left|V_j\right| \left(G_{ij} \cos \delta_{ij} + B_{ij} \sin \delta_{ij}\right) \approx + |V_i| \left|V_j\right| G_{ij}, \tag{7.23}$$
$$\mathcal{Q}_{ij}(\mathbf{x}) = |V_i| \left|V_j\right| \left(G_{ij} \sin \delta_{ij} - B_{ij} \cos \delta_{ij}\right) \approx - |V_i| \left|V_j\right| B_{ij}. \tag{7.24}$$

Note that for $i = j$ these approximations are exact, since $\delta_{ii} = 0$.

From assumption (7.22) it then follows that

$$\left|G_{ij}\right| \approx \left|\mathcal{P}_{ij}(\mathbf{x})\right| \ll \left|\mathcal{Q}_{ij}(\mathbf{x})\right| \approx \left|B_{ij}\right|. \tag{7.25}$$

This leads to the idea of decoupling, i.e., neglecting the off-diagonal blocks of the Jacobian matrix, which are based on G_{ij} and $\mathcal{P}_{ij}$, as they are small compared to the diagonal blocks, which are based on B_{ij} and $\mathcal{Q}_{ij}$.

By the above assumptions, the first order derivatives that constitute the Jacobian matrix of the Newton power flow process can be approximated as follows. Note that assumption (7.21) is used in the first two equations, though only on $\left|V_j\right|$, and that it is assumed that $i \neq j$ whenever applicable.

$$\frac{\partial P_i}{\partial \delta_j}(\mathbf{x}) = \mathcal{Q}_{ij}(\mathbf{x}) \approx -|V_i|\left|V_j\right| B_{ij} \approx -|V_i| B_{ij}, \tag{7.26}$$

$$\frac{\partial P_i}{\partial \delta_i}(\mathbf{x}) = -\sum_{k \neq i} \mathcal{Q}_{ik}(\mathbf{x}) \approx \sum_{k \neq i} |V_i|\,|V_k|\, B_{ik} \approx |V_i| \sum_{k \neq i} B_{ik}, \tag{7.27}$$

$$\frac{\partial Q_i}{\partial \delta_j}(\mathbf{x}) = -\mathcal{P}_{ij}(\mathbf{x}) \approx 0, \tag{7.28}$$

$$\frac{\partial Q_i}{\partial \delta_i}(\mathbf{x}) = \sum_{k \neq i} \mathcal{P}_{ik}(\mathbf{x}) \approx 0, \tag{7.29}$$

$$\frac{\partial P_i}{\partial \left|V_j\right|}(\mathbf{x}) = \frac{\mathcal{P}_{ij}(\mathbf{x})}{\left|V_j\right|} \approx 0, \tag{7.30}$$

$$\frac{\partial P_i}{\partial |V_i|}(\mathbf{x}) = 2\,|V_i|\, G_{ii} + \sum_{k \neq i} \frac{\mathcal{P}_{ik}(\mathbf{x})}{|V_i|} \approx 0, \tag{7.31}$$

$$\frac{\partial Q_i}{\partial \left|V_j\right|}(\mathbf{x}) = \frac{\mathcal{Q}_{ij}(\mathbf{x})}{\left|V_j\right|} \approx -|V_i|\, B_{ij}, \tag{7.32}$$

$$\frac{\partial Q_i}{\partial |V_i|}(\mathbf{x}) = -2\,|V_i|\, B_{ii} + \sum_{k \neq i} \frac{\mathcal{Q}_{ik}(\mathbf{x})}{|V_i|} \approx -2\,|V_i|\, B_{ii} - \sum_{k \neq i} |V_k|\, B_{ik}. \tag{7.33}$$

The last equation (7.33) still requires some work. To this purpose, define the negated row sum $\mathcal{D}_i$ of the imaginary part B of the admittance matrix:

$$\mathcal{D}_i = \sum_{k} -B_{ik} = -B_{ii} - \sum_{k \neq i} B_{ik}. \tag{7.34}$$

Note that, if the diagonal elements of B are negative and the off-diagonal elements are nonnegative, then $\mathcal{D}_i$ is the diagonal dominance of row i. In a system with only generators, loads, and transmission lines without line shunts, $\mathcal{D}_i = 0$ for all i.

Using assumption (7.21) on Eq. (7.33) to approximate $|V_k|$ by $|V_i|$ gives

$$\begin{aligned}
\frac{\partial Q_i}{\partial |V_i|}(\mathbf{x}) &\approx -2\,|V_i|\, B_{ii} - \sum_{k \neq i} |V_k|\, B_{ik} \\
&\approx -2\,|V_i|\, B_{ii} - |V_i| \sum_{k \neq i} B_{ik} \\
&= |V_i| \sum_{k \neq i} B_{ik} - 2\,|V_i| \left(B_{ii} + \sum_{k \neq i} B_{ik} \right) \\
&= |V_i| \sum_{k \neq i} B_{ik} + 2\,|V_i|\, \mathcal{D}_i.
\end{aligned} \tag{7.35}$$

The only term left, in the approximated Jacobian matrix, that depends on the current iterate, now is $|V_i|$. Because of assumption (7.21) this term can be simply set to 1. Another common strategy to remove the dependence on the current iterate from the approximated Jacobian matrix, is to divide each linear equation i by $|V_i|$ in every iteration of the FDLF process. In both cases, the coefficient matrices are the same and constant throughout all iterations. The off-diagonal blocks of these matrices are 0. The upper and lower diagonal blocks are referred to as B' and B'' respectively:

$$B'_{ij} = -B_{ij} \ (i \neq j), \tag{7.36}$$

$$B'_{ii} = \sum_{k \neq i} B_{ik}, \tag{7.37}$$

$$B''_{ij} = -B_{ij} \ (i \neq j), \tag{7.38}$$

$$B''_{ii} = \sum_{k \neq i} B_{ik} + 2\mathcal{D}_i. \tag{7.39}$$

Note that, in a system with only generators, loads and transmission lines, B' is equal to $-B$ without any line shunts incorporated, while B'' is equal to $-B$ with double line shunt values.

Summarising, the FDLF method calculates the iterate update in iteration k by solving the following linear systems:

$$B' \Delta\delta^k = \Delta\mathbf{P}^k, \tag{7.40}$$

$$B'' \Delta|\mathbf{V}|^k = \Delta\mathbf{Q}^k, \tag{7.41}$$

with either

$$\Delta P_i^k = P_i - P_i\left(\delta^k, |\mathbf{V}|^k\right) \text{ and } \Delta Q_i^k = Q_i - Q_i\left(\delta^k, |\mathbf{V}|^k\right), \tag{7.42}$$

or

$$\Delta P_i^k = \frac{P_i - P_i(\delta^k, |\mathbf{V}|^k)}{|V_i|} \text{ and } \Delta Q_i^k = \frac{Q_i - Q_i(\delta^k, |\mathbf{V}|^k)}{|V_i|}. \tag{7.43}$$

7.2.2 Shunts and Transformers

A few additional notes can be made with respect to shunts and transformers, the treatment of which is different between B' and B''.

Shunts have the same influence on the system as transmission line shunts, i.e., they only change the diagonal entries of the admittance matrix. Thus, shunts are left out in B', and doubled in B''.

The modulus $|T|$ of the transformer ratio changes the voltage magnitude, and is therefore generally set to 1 in the calculation of B', which works on the voltage phase angle. Likewise, the argument arg (T) changes the voltage phase angle, and is usually set to 0 for the calculation of B'', which works on the voltage magnitude.

7.2.3 BB, XB, BX, and XX

The Fast Decoupled Load Flow method derived in Sect. 7.2.1 is referred to as the BB scheme, because the susceptance values

$$B_{ij} = \text{Im}\left(\frac{1}{R_{ij} + \iota X_{ij}}\right) = \frac{-X_{ij}}{R_{ij}^2 + X_{ij}^2}. \tag{7.44}$$

are used for both B' and B''.

Stott and Alsac [3] already reported improved convergence in many power flow problems, if the series resistance R was neglected in B', i.e., if for B' the values

$$B_{ij}^X = \text{Im}\left(\frac{1}{\iota X_{ij}}\right) = \frac{-1}{X_{ij}} \tag{7.45}$$

are used instead of the full susceptance. This method is called the XB scheme, because B' is derived from the reactance values X_{ij}, and B'' from the susceptance values B_{ij}.

Van Amerongen [4] found that the BX scheme, where B' is derived from the susceptance values B_{ij}, and B'' from the reactance values X_{ij}, yields convergence that is comparable to XB in most cases, and considerably better in some. Further, he noted that an XX scheme is never better than the BX and XB schemes.

Monticelli et al. [5] presented mathematical support for the good results obtained with the XB and BX schemes. Their idea is the following. Starting with assumptions (7.20) and (7.21), the Jacobian system of the Newton power flow method can be approximated by

$$\begin{bmatrix} -B & G \\ -G & -B \end{bmatrix} \begin{bmatrix} \Delta\delta \\ \Delta|\mathbf{V}| \end{bmatrix} = \begin{bmatrix} \Delta\mathbf{P} \\ \Delta\mathbf{Q} \end{bmatrix}. \tag{7.46}$$

For simplicity, the differences between the diagonals of the upper-left and lower-right blocks, as well as those of the lower-left and upper-right blocks, are neglected.

It should be noted, that the remarks on the incorporation of line shunts described in Sect. 7.2.1, and those on shunts and transformers described in Sect. 7.2.2, remain useful to improve convergence.

Using downward block Gaussian elimination on the system (7.46) gives

$$\begin{bmatrix} -B & G \\ 0 & -\left(B + GB^{-1}G\right) \end{bmatrix} \begin{bmatrix} \Delta\delta \\ \Delta|\mathbf{V}| \end{bmatrix} = \begin{bmatrix} \Delta\mathbf{P} \\ \Delta\mathbf{Q} - GB^{-1}\Delta\mathbf{P} \end{bmatrix}. \tag{7.47}$$

This linear system is solved in three steps, that are then combined into the two steps of the BX scheme.

Step 1: Calculate the partial voltage angle update $\Delta\delta_B^k$ from

$$-B\Delta\delta_B^k = \Delta\mathbf{P}\left(\delta^k, |\mathbf{V}|^k\right) \Rightarrow \Delta\delta_B^k = -B^{-1}\Delta\mathbf{P}\left((\delta)^k, |\mathbf{V}|^k\right), \tag{7.48}$$

where k is the current FDLF iteration.

Step 2: Calculate the voltage magnitude update $\Delta|\mathbf{V}|^k$ from

$$-\left(B + GB^{-1}G\right)\Delta|\mathbf{V}|^k \approx \Delta\mathbf{Q}\left(\delta^k + \Delta\delta_B^k, |\mathbf{V}|^k\right). \tag{7.49}$$

This is an approximation of the lower block of linear equations in (7.47), since the first order Taylor expansion can be used to write

$$\begin{aligned}\Delta\mathbf{Q}\left(\delta^k + \Delta\delta_B^k, |\mathbf{V}|^k\right) &\approx \Delta\mathbf{Q}\left(\delta^k, |\mathbf{V}|^k\right) + \frac{\partial\Delta\mathbf{Q}}{\partial\delta}\left(\delta^k, |\mathbf{V}|^k\right)\Delta\delta_B^k \\ &\approx \Delta\mathbf{Q}\left(\delta^k, |\mathbf{V}|^k\right) - GB^{-1}\Delta\mathbf{P}\left(\delta^k, |\mathbf{V}|^k\right).\end{aligned} \tag{7.50}$$

Here it is used that the partial derivative of $\Delta\mathbf{Q}$ with respect to δ is in the bottom-left block of the Jacobian matrix (7.11), which is approximated by the matrix $-G$ in accordance with Eq. (7.46).

Step 3: Calculate a second partial voltage angle update $\Delta\delta_G^k$ from

$$B\Delta\delta_G^k = G\Delta|\mathbf{V}|^k \Rightarrow \Delta\delta_G^k = B^{-1}G\Delta|\mathbf{V}|^k. \tag{7.51}$$

Then the solution of the upper block of equations in (7.47) is given by

$$\Delta\delta^k = -B^{-1}\Delta\mathbf{P}\left(\delta^k, |\mathbf{V}|^k\right) + B^{-1}G\Delta|\mathbf{V}|^k = \Delta\delta_B^k + \Delta\delta_G^k. \tag{7.52}$$

The next step would be step 1 of the next iteration, i.e.,

$$\Delta\delta_B^{k+1} = -B^{-1}\Delta\mathbf{P}\left(\delta^{k+1}, |\mathbf{V}|^{k+1}\right). \tag{7.53}$$

However, note that

$$\Delta\delta_B^{k+1} + \Delta\delta_G^k \tag{7.54}$$

$$= -B^{-1}\Delta\mathbf{P}\left(\delta^{k+1}, |\mathbf{V}|^{k+1}\right) + B^{-1}G\Delta|\mathbf{V}|^k \tag{7.55}$$

$$= -B^{-1}\left(\Delta\mathbf{P}\left(\delta^k + \Delta\delta_B^k + \Delta\delta_G^k, |\mathbf{V}|^{k+1}\right) - G\Delta|\mathbf{V}|^k\right) \tag{7.56}$$

$$\approx -B^{-1}\left(\Delta\mathbf{P}\left(\delta^k + \Delta\delta_B^k, |\mathbf{V}|^{\mathbf{k+1}}\right) + B\Delta\delta_G^k - G\Delta|\mathbf{V}|^k\right) \tag{7.57}$$

$$= -B^{-1}\Delta\mathbf{P}\left(\delta^k + \Delta\delta_B^k, |\mathbf{V}|^{k+1}\right), \tag{7.58}$$

where a first order Taylor expansion, similar to the one in Eq. (7.50), is used to go from Eqs. (7.56) to (7.57). Thus, instead of calculating δ_G^k to update the voltage angle with it, and then calculating δ_B^{k+1} from Eq. (7.53) to again update the voltage angle, instead a single combined voltage angle update $\Delta\delta_B^{k+1} + \Delta\delta_G^k$ can be calculated from Eq. (7.58).

The above observations lead to the following iteration scheme:

solve $-B\Delta\delta = \Delta\mathbf{P}(\delta, |\mathbf{V}|)$,
update $\delta := \delta + \Delta\delta$,
solve $-\left(B + GB^{-1}G\right)\Delta|\mathbf{V}| = \Delta\mathbf{Q}(\delta, |\mathbf{V}|)$,
update $|\mathbf{V}| := |\mathbf{V}| + \Delta|\mathbf{V}|$.

Note that $\Delta\delta$ here denotes the combined update from Eq. (7.54).

It thus remains to show that the matrix $-\left(B + GB^{-1}G\right)$ is properly represented in the FDLF method.

To this purpose, write $B = A^T dBAy$ and $G = A^T dGA$, where A is the incidence matrix of the associated graph (see Sect. 2.4) and the matrices dB and dG are the diagonal matrices of edge susceptances and edge conductances respectively.

There are two special cases in which this notation can be used to simplify the matrix $-\left(B + GB^{-1}G\right)$. First, if the network is radial then A can be set up as a square nonsingular matrix, see [5], and

$$\begin{aligned}
-\left(B + GB^{-1}G\right) &= -A^T dBA - A^T dGA\left(A^T dBA\right)^{-1} A^T dGA \\
&= -A^T dBA - A^T dGAA^{-1} dB^{-1} A^{-T} A^T dGA \\
&= -A^T dBA - A^T\left(dG^2 dB^{-1}\right) A.
\end{aligned} \tag{7.59}$$

For the second case, note that

$$B_{ij} = \frac{-X_{ij}}{R_{ij}^2 + X_{ij}^2} = -\frac{X_{ij}}{R_{ij}}\frac{R_{ij}}{R_{ij}^2 + X_{ij}^2} = -\frac{X_{ij}}{R_{ij}}G_{ij}. \tag{7.60}$$

Therefore, if the R/X ratio $\rho = \frac{R_{ij}}{X_{ij}}$ is equal on all branches of the power system, then $dB = -\frac{1}{\rho}dG$, and

$$\begin{aligned}
-\left(B+GB^{-1}G\right) &= -A^T dBA - A^T dGA\left(A^T dBA\right)^{-1} A^T dGA \\
&= -A^T dBA + \rho A^T dGA\left(A^T dGA\right)^{-1} A^T dGA \\
&= -A^T dBA + \rho A^T dGA \\
&= -A^T dBA - A^T\left(dG^2 dB^{-1}\right) A.
\end{aligned} \tag{7.61}$$

Both cases lead to the same result, which can be further simplified to

$$\begin{aligned}
-\left(B+GB^{-1}G\right) &= -A^T dBA - A^T\left(dG^2 dB^{-1}\right) A \\
&= -A^T\left(dB^2 + dG^2\right) dB^{-1} A \\
&= -A^T\left(dX^{-1}\right) A,
\end{aligned} \tag{7.62}$$

where $A^T\left(dX^{-1}\right)A$ is equal to the matrix B^X, as defined in Eq. (7.45).

For general networks, if the R/X ratios do not vary a lot, the matrix constructed from the inverse reactances X_{ij}^{-1} can therefore be used as an approximation of the Schur complement matrix $\left(B+GB^{-1}G\right)$. This leads to the BX scheme of the Fast Decoupled Load Flow method.

Similar to the above derivation, starting with the linear system (7.46), and applying block Gaussian elimination upward instead of downward, the XB scheme can be derived. However, when there are PV buses, the convergence of this scheme becomes less reliable than that of the BX scheme. This can be understood by analysing what happens to the XB scheme if all buses are PV buses. In this case the vector $|\mathbf{V}|$ is known, and the linear system from Eq. (7.46) reduces to

$$-B\Delta\delta = \Delta\mathbf{P}. \tag{7.63}$$

In the BX scheme, this is indeed the system that is solved. However, in the XB scheme the coefficient matrix $-B^X$ is used instead of $-B$, leading to unnecessary extra approximation errors.

Summarising, with the assumptions that $\delta_{ij} \approx 0$ and $|V_i| \approx 1$, and the assumption that the R/X ratio does not vary too much between different branches in the network, the BX and XB schemes of the Fast Decoupled Load Flow method can be derived. The assumption on the R/X ratios replaces the original assumption (7.22). The BX and XB schemes of the Fast Decoupled Load Flow method are not decoupled in the original meaning of the term, because the off-diagonal blocks are not disregarded, but are incorporated in the method. As such, these schemes generally have better convergence properties than the BB scheme.

7.3 Convergence and Computational Properties

The convergence of Newton power flow is generally better than that of Fast Decoupled Load Flow, since the FDLF method is an approximation of Newton power flow. The Newton-Raphson method has quadratic convergence when the iterate is close enough to the solution. Fast Decoupled Load Flow often exhibits convergence that is approximately linear. FDLF convergence may be close to the Newton power flow convergence in early iterations, when the iterate is relatively far from the solution. But when the iterate is closer to the solution, Newton power flow converges much faster. Furthermore, in some cases FDLF may fail to converge, while Newton power flow can still find a solution.

Newton power flow and the Fast Decoupled Load Flow method both evaluate the power mismatch function in every iteration. The FDLF method calculates the coefficient matrices B' and B'' only once at the start. In the case of Newton power flow, the Jacobian matrix has to be calculated in every iteration. However, the Jacobian matrix can be computed at relatively little extra cost when evaluating the power mismatch function, as was shown in Sect. 7.1.2. Thus, there is no significant computational difference in terms of the evaluation of the power mismatch and coefficient matrices.

Both algorithms traditionally use a direct method to solve the linear systems of equations. Newton power flow needs to make an LU decomposition of the Jacobian in each iteration. In case of the FDLF method, the LU decomposition of B' and B'' can be made once at the start. Then, in every iteration, only forward and backward substitutions are needed to solve the linear systems, reducing computational cost (see Sect. 3.1.3). Furthermore, the FDLF coefficient matrices B' and B'' each hold about a quarter of the number of nonzeros that the Jacobian matrix has, reducing memory requirements and computational cost compared to Newton power flow.

Summarising, the choice between Newton power flow and Fast Decoupled Load Flow is about reducing computational and memory cost per iteration, at the cost of convergence speed and robustness.

In practice, Newton power flow is usually preferred over FDLF because of the improved robustness. In the discussion of [6], it was also agreed upon that for the large complex power flow problems of the future, the focus should be on Newton power flow, rather than Fast Decoupled Load Flow. As discussed in the remainder of this book, both in theory and experiments, Newton-Krylov power flow methods offer the best of both.

7.4 Interpretation as Elementary Newton–Krylov Methods

Both traditional Newton power flow and Fast Decoupled Load Flow can be seen as simple Newton-Krylov power flow solvers, that perform a single Richardson iteration in each Newton step. In the case of Newton power flow the Richardson iteration is preconditioned using an LU factorisation of the Jacobian matrix. In the case of Fast

Decoupled Load Flow, the preconditioner instead is an FDLF operator using LU factorisations of B' and B''.

This interpretation shows a clear path towards improving the traditional power flow solvers. The single Richardson iteration can be replaced by the combination of a more efficient Krylov method, like GMRES, Bi-CGSTAB, or IDR(s), and a good strategy for choosing the forcing terms.

For Fast Decoupled Load Flow this leads directly to a proper Newton-Krylov method, preconditioned with the FDLF operator. Provided that the used Krylov method converges linearly or better, the total amount of linear iterations performed is no larger than the total amount of Richardson iterations needed for FDLF (see Chap. 5), while the amount of nonlinear iterations goes down, and the convergence and robustness improve to the level of Newton power flow.

Newton power flow needs some more work. Since the preconditioner is a direct solve on the coefficient matrix, a single linear iteration leads to convergence independent of the Krylov method and the forcing terms. Thus the preconditioner has to be relaxed. Obvious candidates are using an incomplete LU factorisation of the Jacobian matrix in each Newton iteration, or a single LU or ILU factorisation of the initial Jacobian J_0 throughout all Newton iterations. A relaxed preconditioner leads to more linear iterations being needed. However, if the calculation of the relaxed preconditioner is sufficiently faster than the direct solves of the traditional method, the overall method will be faster.

In Chap. 8 we investigate the use of Newton-Krylov solvers for power flow problems in detail, and compare the performance of these methods with that of a traditional Newton power flow implementation.

References

1. Tinney, W.F., Hart, C.E.: Power flow solution by Newton's method. IEEE Trans. Power Apparatus Syst. **86**(11), 1449–1449 (1967)
2. Tinney, W.F., Walker, J.W.: Direct solutions of sparse network equations by optimally ordered triangular factorization. Proc. IEEE **55**(11), 1801–1809 (1967)
3. Stott, B., Alsac, O.: Fast decoupled load flow. IEEE Trans. Power Apparatus Syst. **93**(3), 859–869 (1974)
4. van Amerongen, R.A.M.: A general-purpose version of the fast decoupled loadflow. IEEE Trans. Power Syst. **4**(2), 760–770 (1989)
5. Monticelli, A.J., Garcia, A., Saavedra, O.R.: Fast decoupled load flow: hypothesis, derivations, and testing. IEEE Trans. Power Syst. **5**(4), 1425–1431 (1990)
6. Dag, H., Semlyen, A.: A new preconditioned conjugate gradient power flow. IEEE Trans. Power Syst. **18**(4), 1248–1255 (2003)

Chapter 8
Newton–Krylov Power Flow Solver

Newton power flow solvers traditionally use a direct method to solve the linear systems [1, 2] (see also Chap. 7). For large linear systems of equations with a sparse coefficient matrix, iterative linear solvers are generally more efficient than direct solvers (see Chap. 3). Using a Krylov method to solve the Jacobian systems in the Newton-Raphson method, leads to a Newton-Krylov method (see Chap. 4). It has been recognised that such solvers offer many advantages over the traditional implementation for large power systems [3–10].

In this chapter, computational aspects of Newton-Krylov power flow solvers are discussed, using the numerical experiments presented in Chap. 10 as a reference. Sect. 8.1 focuses on which Krylov method to use, followed by a discussion on preconditioning in Sect. 8.2, and different forcing term strategies in Sect. 8.3. Then, Sect. 8.4 gives an overview of the speed and scaling of Newton-Krylov power flow solvers, and Sect. 8.5 discusses their robustness.

We show that direct solvers, and other methods using the LU factorisation, scale very badly in the problem size. The alternatives proposed in this chapter are faster for all tested problems and have near linear scaling, thus being much faster for large power flow problems. The largest test problem, with a million buses, takes over an hour to solve using a direct solver, while a Newton-Krylov solver can solve it in less than 30 seconds. That is 120 times faster.

8.1 Linear Solver

In every Newton iteration, a linear system has to be solved of the form

$$-J_i\mathbf{s}_i = \mathbf{F}_i. \tag{8.1}$$

For a power flow problem, the Jacobian matrix J_i can be calculated at very little extra cost when evaluating the power mismatch $\mathbf{F}_i$ (see Sect. 7.1.2). Therefore, there is no need to resort to an approximate Jacobian, or Jacobian-free method.

R. Idema and D. J. P. Lahaye, *Computational Methods in Power System Analysis*,
Atlantis Studies in Scientific Computing in Electromagnetics,
DOI: 10.2991/978-94-6239-064-5_8, © Atlantis Press and the authors 2014

The linear solver considered in this chapter is preconditioned GMRES. GMRES has the optimality property, and thus solves the linear problem in the minimal number of iterations generally needed by a Krylov method. The number of iterations needed to converge says something about the quality of the preconditioner. Whether restarted GMRES, or other Krylov methods, like Bi-CGSTAB or IDR(s), should also be considered, can be derived from the performance of GMRES.

In our experiments, the best results are obtained with high quality preconditioners. Then, only a low number of GMRES iterations is needed per Newton iteration. At such low iteration counts there is no reason to restart GMRES, nor to drop the optimality property for a method with short recurrences. Bi-CGSTAB is faster than GMRES for lower quality preconditioners; however, GMRES with a high quality preconditioner is still faster than Bi-CGSTAB with lower quality preconditioners. IDR(s) should lead to similar results as Bi-CGSTAB.

8.2 Preconditioning

Preconditioning is essential to the performance of Krylov methods such as GMRES. The presented Newton-Krylov power flow solvers use right preconditioning. This means that linear systems of the form

$$J_i P_i^{-1} \mathbf{z}_i = -\mathbf{F}_i \tag{8.2}$$

are solved, and that the Newton step $\mathbf{s}_i$ follows from solving $P_i \mathbf{s}_i = \mathbf{z}_i$. To achieve fast convergence, the preconditioner matrix P_i should resemble the coefficient matrix J_i. At the same time, a fast way to solve linear systems of the form $P_i \mathbf{u}_i = \mathbf{v}_i$ is needed, as such a system has to be solved in each iteration of the linear solver.

In this book we consider preconditioners in the form of a product of a lower and upper triangular matrix $P_i = L_i U_i$. Any linear system with coefficient matrix P_i can then simply be solved using forward and backward substitution. To get such a preconditioner, we choose a target matrix Q_i that resembles J_i, and construct the LU or an ILU(k) factorisation.

Other preconditioners that are known to often work well for large problems are preconditioners based on iterative methods. Only stationary iterative methods can be used as a preconditioner for standard implementations of GMRES, Bi-CGSTAB, and IDR(s). Nonstationary iterative methods, like GMRES itself, can only be used with special flexible iterative methods, like FGMRES [11]. The use of FGMRES with a GMRES based preconditioner was explored in [8, 10].

Algebraic Multigrid (AMG) methods can also be used as preconditioner. A cycle of AMG, with a stationary solver on the coarsest grid, leads to a stationary preconditioner. Such a preconditioner is very well suited for extremely large problems. For more information on AMG see [12, App. A]. The use of AMG as a preconditioner for Newton-Krylov power flow was explored in [10].

8.2.1 Target Matrices

The target matrix Q_i is the matrix from which the preconditioner is derived. There are three common choices for the target matrix: the Jacobian matrix $Q_i = J_i$, the initial Jacobian matrix $Q_i = J_0$, and $Q_i = \Phi$, where

$$\Phi = \begin{bmatrix} B' & 0 \\ 0 & B'' \end{bmatrix}, \tag{8.3}$$

with B' and B'' as in the BX scheme of the Fast Decoupled Load Flow method.

The FDLF matrix Φ can be seen as an approximate Schur complement of the initial Jacobian matrix, as discussed in Sect. 7.2.3. This matrix has been shown to be a good preconditioner [5, 9, 10]. It is a lower quality preconditioner than the initial Jacobian matrix itself; however, it consists of two decoupled blocks with each about a quarter of the nonzeros found in the Jacobian matrix. The structure, and the lower nonzero count, provide benefits for computing time and memory usage.

8.2.2 Factorisation

The used preconditioners are LU and ILU(k) factorisations of the target matrices. For the LU factorisation this leads to preconditioners $P_i = L_i U_i = Q_i$, whereas with ILU factorisations the preconditioners $P_i = L_i U_i$ only resemble the target matrix Q_i. An ILU factorisation is generally cheaper to build than the LU factorisation, whereas the LU factorisation results in a higher quality preconditioner.

The quality of the LU preconditioner is predetermined by the target matrix, since $P_i = Q_i$. The LU decomposition of a sparse matrix generally leads to a certain amount of fill-in (see Sect. 3.1.4). The more the fill-in, the more memory and computational time are needed for the factorisation and the forward and backward substitutions. It is therefore important to minimise the fill-in of the LU factorisation, by choosing a proper ordering of the rows and columns of the target matrix Q_i.

With the ILU(k) factorisation, the fill-in ratio is influenced both by the number of levels k, and the row and column ordering. The effect of the matrix ordering on the fill-in is much less pronounced than with the LU decomposition; however, the ordering also influences the quality with which the ILU factorisation approximates the target matrix Q_i (see Sect. 3.1.5).

Experiments show that the Approximate Minimum Degree (AMD) [13] reordering gives the best results for power flow problems. The AMD ordering takes relatively little time to compute, while yielding the least fill-in with a full LU factorisation and the best quality ILU factorisations at the same time. See Sect. 10.1 for an overview of experiments that inspire these conclusions.

8.2.3 Reactive Power Limits and Tap Changing

For a practical power flow solver, it is very important to be able to efficiently deal with solution adjustments due to reactive power limits of generators or to transformer taps. Generally speaking, all solution adjustments that can be handled using a direct solver, can also be solved effectively using preconditioned Krylov methods by making a new preconditioner based on the Jacobian in every iteration. However, it is often more efficient to keep the preconditioner constant over multiple Newton iterations, as is also clear from the numerical experiments in Sect. 10.3. Below, the options for keeping the preconditioner constant while applying these solution adjustments, are discussed.

Reactive power limits of generators are usually handled by checking for violations during the Newton iterations. When a violation is detected, the representation of the corresponding generator is changed from a PV bus to a PQ bus with reactive power Q equal to the violated bound.

If the power flow method is implemented to eliminate reactive power equations of PV buses, then this bus-type switching changes the dimensions of the linear system (see Sect. 7.1.3). It is not directly clear how to reuse the preconditioner then. However, there are several ways to implement a power flow solver that keeps the dimensions constant. The reactive power equations of PV buses can be kept, but with very large values on the corresponding diagonal entries of the Jacobian, as also described in [14]. This may negatively impact the condition number of the Jacobian. A better method, in the context of preconditioned Krylov methods, is to keep the diagonal elements as-is, but set the other entries for those equations to 0, including the right-hand side value. The resulting equations are very simple to deal with for factorisation methods, leading to hardly any increase of computational effort compared to elimination.

Tap changing transformers can be dealt with in two ways: automatic adjustment or error-feedback adjustment. Automatic adjustment methods change the power flow equations to incorporate the tap settings as variables [15]. As a result, the structure of the Jacobian changes, and it is not clear how to use an FDLF based matrix as preconditioner. However, preconditioners based on the Jacobian itself can be applied as normal. Error-feedback adjustments do not change the structure of the Jacobian, but adjust the equations between Newton iterations [16]. As such, all the discussed methods of preconditioning can be used.

8.3 Forcing Terms

Inexact Newton methods solve the Jacobian system in iteration i such that

$$\|\mathbf{F}(\mathbf{x}_i) + J(\mathbf{x}_i)\mathbf{s}_i\| \leq \eta_i \|\mathbf{F}(\mathbf{x}_i)\|, \tag{8.4}$$

where η_i are called the forcing terms. Choosing the forcing terms correctly is very important, as discussed in detail in Chap. 5. Below, two strategies for choosing the forcing terms are discussed.

The first strategy is based on work by Dembo and Steihaug [17]:

$$\eta_i = \min\left\{\frac{1}{2^i}, \|\mathbf{F}_i\|\right\}. \tag{8.5}$$

This allows for superlinear convergence when the iterate is far from the solution, where $\frac{1}{2^i}$ will be the smaller term. When nearing the solution, where $\|\mathbf{F}_i\|$ is smaller, this method switches to quadratic convergence.

The second strategy is by Eisenstat and Walker [18]. This method starts with some choice of initial forcing term η_0, and for $i > 0$ sets

$$\eta_i = \left|\frac{\|\mathbf{F}_i\| - \|\mathbf{F}_{i-1} + J_{i-1}\mathbf{s}_{i-1}\|}{\|\mathbf{F}_{i-1}\|}\right|, \tag{8.6}$$

while safeguarding from oversolving by adding the rule

$$\text{if } \eta_{i-1}^{\frac{1}{2}+\frac{1}{2}\sqrt{5}} > \frac{1}{10}, \text{ then } \eta_i = \max\left\{\eta_i, \eta_{i-1}^{\frac{1}{2}+\frac{1}{2}\sqrt{5}}\right\}. \tag{8.7}$$

In our experiments $\eta_0 = 0.1$ was used.

Some authors have used fixed forcing terms throughout all Newton iterations for power flow problems [3–5, 7]. This is generally not a good idea. If the chosen forcing terms are small, then a lot of oversolving is done in early iterations, leading to many extra GMRES iterations. If the forcing terms are large, then there is a lot of undersolving in later iterations, leading to many extra Newton iterations. And anything in between leads to both oversolving in early iterations, and undersolving in later iterations.

The strategy by Eisenstat and Walker is successfully being used in practice on many different types of problems, as it is able to adapt to the problem very well. It also provides very good results for power flow problems. The method based on the work of Dembo and Steihaug also gives good results, but generally leads to some undersolving for the tested power flow problems. For more details see the experiments and discussion in Sects. 10.2 and 10.3.

Table 8.1 Power flow for small test cases

Problem	uctew001	uctew002	uctew004	uctew008
Direct	0.077	0.16	0.33	0.69
LU of J_0	0.060	0.12	0.25	0.52

8.4 Speed and Scaling

This section gives an overview of the speed and scaling of Newton-Krylov power flow solvers. The numerical experiments on which this section is based can be found in Sect. 10.3. An explanation of the used test cases is provided in Chap. 11.

For the smaller test cases, using the LU factorisation of the J_0 target matrix leads to the best results for both forcing term strategies. The Eisenstat and Walker forcing terms perform slightly better than the alternative by Dembo and Steihaug. Table 8.1 compares the resulting solution times in seconds, with the solution times of the traditional implementation with a direct solver.

Solving a single power flow problem of these small sizes is so fast, that all tested methods are acceptable. However, when using the power flow solver as part of a larger system that has to solve many power flow problems, as for example in contingency analysis, using the LU factorisation of J_0 can lead to a significant reduction of computing time.

For the larger test cases, any method using the LU decomposition is slow due to the bad scaling of that operation (see Sects. 10.1.1 and 10.3.1). The best results are obtained using ILU(12) factorisations of the J_0 and Φ target matrices. Again, the Eisenstat and Walker forcing terms perform slightly better than the Dembo and Steihaug strategy.

Figure 8.1 shows the solution time in seconds for LU and ILU(12) factorisations of the J_0 and Φ target matrices, using Eisenstat and Walker forcing terms. The bad scaling of the LU factorisation is clearly visible. It is worse for J_0 than for the Φ target matrix, because the Φ matrix consists of two independent blocks of half the dimension, see also Sect. 10.1.1.

The ILU(12) factorisation leads to near linear scaling. Which of the target matrices J_0 and Φ leads to the fastest solution differs per test case. Note that this has to do with being a bit lucky, or unlucky, at how the inexact Newton steps turn out exactly for a certain problem and forcing terms, much more than it has to do with the quality of the preconditioner.

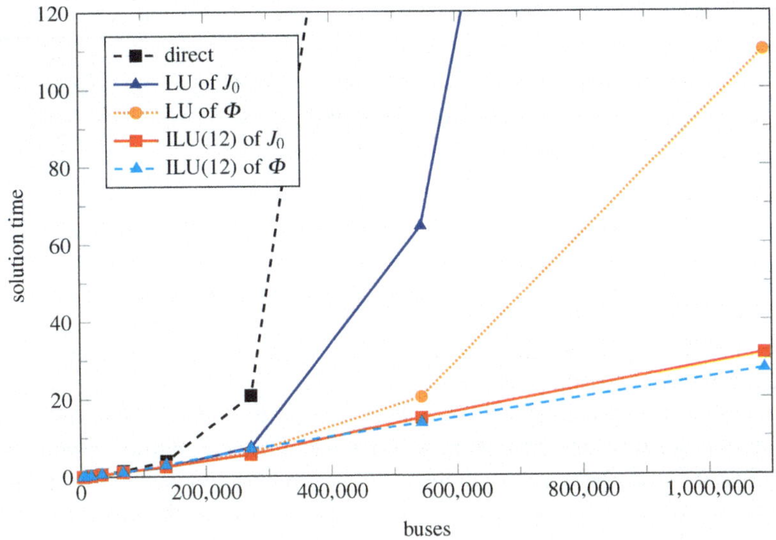

Fig. 8.1 Newton-Krylov power flow for all test cases

8.5 Robustness

This section investigates the robustness of the Newton-Krylov power flow solver, and compares it with that of traditional Newton power flow. Convergence of exact and inexact Newton methods are very close, provided that the forcing terms are chosen small enough. Therefore, to compare the robustness of these methods, the linear solvers should be compared. Direct linear solvers are very robust, thus the question is whether the used iterative linear solver is robust.

The convergence of Krylov methods depends on the Krylov subspace, which is determined by the coefficient matrix, the right-hand side vector, and the initial iterate. The condition number of the coefficient matrix can give some indication on how fast Krylov methods will converge for problems with that coefficient matrix. The Jacobian matrices of our test cases are quite ill-conditioned. For example, the condition number of the initial Jacobian J_0 of the uctew001 test case is 1.2×10^9. This is why preconditioning is such an important part of the solution process. Due to the large condition number, low quality preconditioners lead to slow convergence. However, with the preconditioners P_i based on LU and ILU(12) factorisations, the condition number of the preconditioned coefficient matrix $J_i P_i^{-1}$ drops below 10, leading to very fast convergence.

Table 8.2 shows some test results for experiments related to solver robustness. In these experiments, the uctew032 test case is solved at different loading levels, using the LU factorisation of J_0 as preconditioner. Presented are the number of Newton iterations N, the GMRES iterations (total amount and amount per Newton iteration), and an estimate $\tilde{\sigma}$ of the condition number of the preconditioned coefficient matrix

Table 8.2 Convergence at different loading levels

Load (%)	N	GMRES iterations	$\tilde{\sigma}$
100	6	21 [1,3,2,3,5,7]	3.5
110	6	21 [1,3,2,3,5,7]	3.5
120	6	22 [1,3,2,3,5,8]	4.1
130	7	35 [1,3,2,3,6,7,13]	4.6
140	7	37 [1,3,2,4,5,8,14]	5.4
150	7	35 [1,3,2,4,6,7,12]	6.8
160	7	34 [1,3,2,4,6,5,13]	10.4

in the final Newton iteration. Note that the other preconditioners treated in this book lead to very similar results.

At higher loading levels, the solution of the power flow problem with lies farther away from the flat start than with lower load. Since the preconditioner is based on the Jacobian at the flat start, at high loading levels the Jacobian near the solution also differs more from the preconditioner. As a result the condition number of the preconditioned coefficient matrix in the last Newton iteration goes up with the loading level, as indicated by the values of $\tilde{\sigma}$. However, overall the condition number stays very small and GMRES convergence does not deteriorate visibly. It does take longer to solve the systems with high load, but this is due to Newton convergence suffering, not the convergence of the linear solver, and is the same when using a direct linear solver.

Both the Newton-Raphson method with direct solver and the Newton-Krylov methods are able to solve the problem up to a loading level of 160 %, but fail to converge at 170 %, without using line search or trust region techniques. It should be noted that the solution of the power flow problem at the highest loading levels has such large voltage angles not to be of practical value, indicating that the solvers can easily handle any practical loading levels of the power system.

References

1. Tinney, W.F., Hart, C.E.: Power flow solution by Newton's method. IEEE Trans. Power Apparatus and Syst. PAS **86**(11), 1449–1449 (1967)
2. Tinney, W.F., Walker, J.W.: Direct solutions of sparse network equations by optimally ordered triangular factorization. Proc. IEEE **55**(11), 1801–1809 (1967)
3. Semlyen, A.: Fundamental concepts of a Krylov subspace power flow methodology. IEEE Trans. Power Syst. **11**(3), 1528–1537 (1996)
4. Pai, M.A., Dag, H.: Iterative solver techniques in large scale power system computation. In: Proceedings of the 36th Conference on Decision & Control, pp. 3861–3866 (1997)
5. Flueck, A.J., Chiang, H.D.: Solving the nonlinear power flow equations with an inexact Newton method using GMRES. IEEE Trans. Power Syst. **13**(2), 267–273 (1998)
6. de León, F., Semlyen, A.: Iterative solvers in the Newton power flow problem: preconditioners, inexact solutions and partial Jacobian updates. IEE Proc. Gener. Transm. Distrib **149**(4), 479–484 (2002)
7. Chaniotis, D., Pai, M.A.: A new preconditioning technique for the GMRES algorithm in power flow and $P - V$ curve calculations. Electr. Power Energy Syst. **25**, 239–245 (2003)

8. Zhang, Y.S., Chiang, H.D.: Fast Newton-FGMRES solver for large-scale power flow study. IEEE Trans. Power Syst. **25**(2), 769–776 (2010)
9. Idema, R., Lahaye, D.J.P., Vuik, C., van der Sluis, L.: Scalable Newton-Krylov solver for very large power flow problems. IEEE Trans. Power Syst. **27**(1), 390–396 (2012)
10. Idema, R., Papaefthymiou, G., Lahaye, D.J.P., Vuik, C., van der Sluis, L.: Towards faster solution of large power flow problems. IEEE Trans. Power Syst. **28**(4), 4918–4925 (2013)
11. Saad, Y.: A flexible inner-outer preconditioned GMRES algorithm. SIAM J. Sci. Comput. **14**(2), 461–469 (1993)
12. Trottenberg, U., Oosterlee, C.W., Schüller, A.: Multigrid. Academic Press, New York (2001)
13. Amestoy, P.R., Davis, T.A., Duff, I.S.: An approximate minimum degree ordering algorithm. SIAM J. Matrix Anal. Appl. **17**(4), 886–905 (1996)
14. Chang, S.K., Brandwajn, V.: Adjusted solutions in fast decoupled load flow. IEEE Trans. Power Syst. **3**(2), 726–733 (1988)
15. Peterson, N.M., Meyer, W.S.: Automatic adjustment of transformer and phase-shifter taps in the Newton power flow. IEEE Trans. Power Apparatus Syst. PAS-90(1), 103–108 (1971)
16. Mamandur, K.R.C., Berg, G.J.: Automatic adjustment of generator voltages in Newton-Raphson method of power flow solutions. IEEE Trans. Power Apparatus Syst. PAS-101(6), 1400–1409 (1982)
17. Dembo, R.S., Steihaug, T.: Truncated-Newton algorithms for large-scale unconstrained optimization. Math. Program. **26**, 190–212 (1983)
18. Eisenstat, S.C., Walker, H.F.: Choosing the forcing terms in an inexact Newton method. SIAM J. Sci. Comput. **17**(1), 16–32 (1996)

Chapter 9
Contingency Analysis

Secure operation of a power system requires not only that the system operates within specified system operating conditions, but also that proper operation is maintained when contingencies occur. Contingency analysis simulates credible contingencies to analyse their impact on the operation of the power system.

Contingency analysis consists of three phases: definition, selection, and evaluation. In the definition phase, a list of contingencies with a non-negligible chance of occurring is constructed. These contingencies mostly consist of single or multiple generator and branch outages. In the selection phase, fast approximation techniques are used to cheaply identify contingency cases that will not violate system operation conditions. These contingencies can be eliminated from the list that was made in the definition phase. Finally, in the evaluation phase, for the remaining contingencies the power flow problem is solved, and the solution analysed. For more information, see for example [1].

In this book we focus on the evaluation phase. In particular, we focus on how Newton-Krylov power flow solvers can be used to speed up the calculation of many very similar power flow problems. Speeding up consecutive solves generally involves reusing information from earlier solves. The information that can be reused in contingency analysis, that is not available in traditional Newton power flow, is the preconditioner. We will see that this can lead to a significant reduction of the computational time.

In this chapter, a methodology is presented to speed up power flow calculations for branch outages in contingency analysis. This methodology can be used whenever power flow problems have to be solved that only differ slightly from each other. This includes other types of outages, Monte Carlo simulations, and optimisation problems. Similar techniques were proposed for contingency screening in [2].

9.1 Simulating Branch Outages

A branch outage can be simulated by removing the branch from the power system model, and then solving the associated power flow problem. In contingency analysis many branch outages are simulated, leading to a large amount of power flow

R. Idema and D. J. P. Lahaye, *Computational Methods in Power System Analysis*, Atlantis Studies in Scientific Computing in Electromagnetics,
DOI: 10.2991/978-94-6239-064-5_9,

problems to be solved. Solving all these problems can take a huge computational effort. Therefore, it is important to look at ways to speed up this process, beyond the efforts of speeding up the power flow solver itself. The general methodology here, is to treat the branch outages as variations of the base case, i.e., the case without any outages, instead of treating them as independent problems.

A simple improvement that can be made, is to update the admittance matrix Y, instead of recalculating it for each outage. Let i and j be the buses that the removed branch used to connect. Then only the values Y_{ii}, Y_{ij}, Y_{ji}, and Y_{jj} in the admittance matrix have to be updated.

Since power systems with a branch outage very closely resemble the base case, it seems logical to assume that the solutions of the associated power flow problems are relatively close to the solution of the base case. Therefore, instead of using a flat start, using the solution of the base case as initial iterate for the contingency cases can be expected to significantly reduce the number of Newton iterations needed to reach convergence.

The above two techniques can be equally exploited in classical Newton power flow, Fast Decoupled Load Flow, and Newton-Krylov power flow. Fast Decoupled Load Flow further allows the reuse of the factorisation of the base case coefficient matrices B' and B'' for contingency cases, either through factor updating or compensation techniques [3–5].

Newton-Krylov power flow provides extra options through preconditioning. A single preconditioner, based on the base case, can be used for all contingency cases, eliminating the need to perform a factorisation for each contingency. This preconditioner may generally not be as good as those derived from the contingency cases themselves, but the resulting extra GMRES iterations are relatively cheap. The base case preconditioner can be further improved for the contingency cases, using the same techniques that are used in Fast Decoupled Load Flow to update the coefficient matrices for simulating branch outages.

Knowledge of the speed of convergence of the base case may also be exploited to speed up contingency analysis, if this knowledge can be used to improve the forcing terms for the contingency cases.

Numerical experiments, simulating branch outages using Newton-Krylov power flow, can be found in Sect. 10.4. These experiments focus on reusing a single preconditioner throughout all contingency cases. Analysis of the forcing terms yielded only minor improvements over the Eisenstat and Walker forcing terms, as those forcing terms already adapt to the problem very well.

Figure 9.1 gives an overview of the results for the UCTE winter 2008 study model (uctew001, see Chap. 11), using Eisenstat and Walker forcing terms. It shows the time spent on LU factorisations, forward and backward substitutions, other GMRES operations, the calculation of the Jacobian systems, and other computations.

The left three bars represent methods using a flat start, while the right three bars use the solution of the base case as initial iterate for the contingency cases. The methods used are:

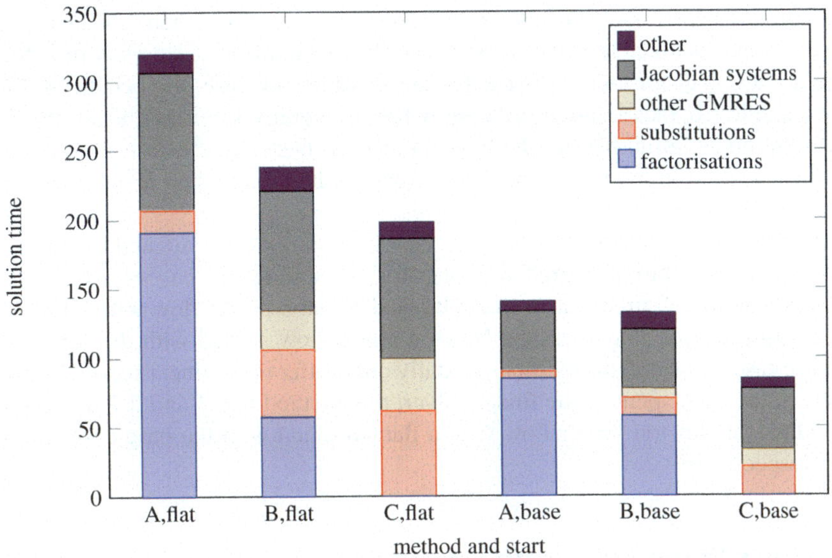

Fig. 9.1 Contingency analysis

A : Newton power flow with a direct linear solver,
B : Newton-GMRES with the contingency cases preconditioned with their own initial Jacobian,
C : Newton-GMRES with the contingency cases preconditioned with the base case Jacobian evaluated in the vector that is used as initial iterate for the contingency cases.

Classical Newton power flow (A) serves as a reference. Method B is chosen because it is the fastest method for solving a single uctew001 case (see Sect. 10.3). Finally, method C illustrates the benefits of using a single preconditioner for all contingency cases.

Starting with the solution of the base case has a clear advantage over a flat start for this test case. Many fewer Newton iterations are needed when using the base case solution as initial iterate, leading to a significant computational speed-up.

With the classical Newton power flow (A), most of the computational effort is spent on the LU factorisations of Jacobian matrices. Such a factorisation has to be made in every Newton iteration. The rest of the linear solving consists of a single forward and backward substitution per Newton iteration. With Newton-GMRES, preconditioned with the initial Jacobian of the case that is being solved (B), only one LU factorisation has to be made per contingency case. This comes at the cost of some extra computational effort for the GMRES iterations (substitutions and other GMRES operations). When preconditioning Newton-GMRES with the base case Jacobian (C), the time spent on LU factorisations becomes negligible, at the cost of some more GMRES iterations.

Method B clearly outperforms classical Newton power flow when using a flat start. The difference is much less distinct when starting from the base case solution. Since there are less Newton iterations needed per contingency case, the benefit of doing only a single LU factorisation per case is less pronounced. Further, using the base case solution as initial iterate, the Newton process is started closer to the solution. As a result the oversolving of the direct solver is less than when using a flat start, thus less effort is wasted.

The best results are obtained using the base case solution as start, and the base case Jacobian in that solution as preconditioner for the contingency cases (C, base). This method is about 3.8 times faster than classical Newton power flow with a flat start, and 1.7 times faster than classical Newton power flow started with the solution of the base case. Note that the methods actually only differ in the linear solver. Looking purely at the time spent in the linear solver, this method is 6.2 and 2.7 times faster than classical Newton power flow with a flat start, and with the base case solution start, respectively.

9.2 Other Simulations with Uncertainty

The methodology described in Sect. 9.1 can also be used in other power flow simulations that include some element of uncertainty, such as optimisation problems and Monte Carlo simulations.

An obvious example is dealing with uncertainty in load. This is a common problem, for example when dealing with wind turbines. Wind turbines are generally modeled as negative loads, and have a high uncertainty in the amount of generated power. Using Monte Carlo simulations to handle these stochastic factors requires the solution of a lot of power flow problems with the same network topology, but different load values. Since the load values do not influence the Jacobian matrix, all these power flow problems have the same initial Jacobian matrix when started with the same initial iterate. As such, that Jacobian matrix can be expected to be a very good preconditioner for all problems within the Monte Carlo simulation.

References

1. Stott, B., Alsac, O., Monticelli, A.J.: Security analysis and optimization. Proc. IEEE **75**(12), 1623–1644 (1987)
2. Alves, A.B., Asada, E.N., Monticelli, A.: Critical evaluation of direct and iterative methods for solving ax = b systems in power flow calculations and contingency analysis. IEEE Trans. Power Syst. **14**(2), 702–708 (1999)
3. Tinney, W.F.: Compensation methods for network solutions by optimally ordered triangular factorization. IEEE Trans. Power Apparatus Syst. PAS **91**(1), 123–127 (1972)
4. Alsac, O., Stott, B., Tinney, W.F.: Sparsity-oriented compensation methods for modified network solutions. IEEE Trans. Power Apparatus Syst. PAS **102**(5), 1050–1060 (1983)
5. van Amerongen, R.A.M.: A rank-oriented setup for the compensation algorithm. IEEE Trans. Power Syst. **5**(1), 283–288 (1990)

Chapter 10
Numerical Experiments

In this chapter, numerical experiments with the Newton-Krylov power flow solver are presented. Many of these experiments have been referenced and summarised in previous chapters. Here, the full experiments are discussed in detail.

Section 10.1 presents numerical experiments with LU and ILU(k) factorisations, testing different matrix ordering methods to determine the most promising options. Section 10.2 analyses some experiments with the different forcing term strategies. Section 10.3 treats experiments with the power flow solver for all combinations of target matrices and forcing terms. Finally, Sect. 10.4 treats experiments regarding contingency analysis.

The Newton-Krylov power flow solver used for the experiments, is implemented in C++ using PETSc (Portable, Extensible Toolkit for Scientific Computation) [1], a state-of-the-art C library for scientific computing. PETSc can be used both to produce sequential programs, and to build parallel code.

All experiments are performed on one core of an Intel Core2 Duo 3 GHz CPU, in a machine with 16 Gb memory, running a 64-bit Slackware 13 Linux distribution.

Table 10.1 shows the dimensions of the used power flow test cases. For detailed information on the test set see Chap. 11. All experiments use a flat start, unless stated otherwise, and all computational times are measured in seconds.

10.1 Factorisation

This section discusses numerical experiments with LU and ILU(k) factorisations, and different row and column ordering methods. In Sect. 10.1.1 different ordering methods are tested for the LU decomposition, and the scaling of this factorisation is investigated. Section 10.1.2 treats the impact of matrix ordering on the ILU(k) factorisation, and tests the speed and scaling for different levels k.

All experiments treated in this section consist of solving the Jacobian system of the first Newton iteration, $J_0\mathbf{s}_0 = \mathbf{F}_0$, using preconditioned GMRES, up to an

R. Idema and D. J. P. Lahaye, *Computational Methods in Power System Analysis*,
Atlantis Studies in Scientific Computing in Electromagnetics,
DOI: 10.2991/978-94-6239-064-5_10,

Table 10.1 Power flow test cases for numerical experiments

Name	Buses	Branches	nnz (J)
uctew001	4,253	7,191	62,654
uctew002	8,505	14,390	125,372
uctew004	17,009	28,796	250,872
uctew008	34,017	57,624	502,000
uctew016	68,033	115,312	1,004,512
uctew032	136,065	230,752	2,010,048
uctew064	272,129	461,760	4,022,144
uctew128	544,257	924,032	8,048,384
uctew256	1,088,513	1,849,088	16,104,960

Table 10.2 PETSc functions concerning factorisation

MatGetOrdering	: Matrix reordering
MatLUFactorSym	: Symbolic LU factorisation
MatILUFactorSym	: Symbolic ILU factorisation
MatLUFactorNum	: Numeric factorisation for LU or ILU
PCSetUp	: Reordering and factorisation
PCApply	: Forward and backward substitution
KSPSolve	: Linear solve

accuracy of 10^{-5}. The experiments are conducted on the uctew032 problem, but similar results hold for the other test cases. In the scaling experiments, the entire range of test cases is used.

The discussion of the experiments includes the computation time spent on the relevant PETSc functions. Table 10.2 gives an overview of these PETSc functions. Further, note that the notation nnz $(L + U)$ is used for the number of nonzeros in the factors L and U combined, and the term fill ratio is used for the number of nonzeros in the factors divided by the number of nonzeros in the target matrix.

10.1.1 LU Factorisation

Table 10.3 shows the results of the direct solution of the first Jacobian system of the uctew032 test case. The direct solver provided in PETSc is tested together with the following matrix ordering methods: Nested Dissection (ND), One-way Dissection (1WD), Reverse Cuthill-McKee (RCM), Quotient Minimum Degree (QMD), and Approximate Minimum Degree (AMD) [2–4]. PETSc is further used to call the UMFPACK [5], MUMPS [6], SuperLU [7], and SuperLU_Dist [8] direct solvers, with the provided ordering methods.

Note that with SuperLU and SuperLU_Dist the natural row ordering is used; the alternative yields no improvement here. Further note that PETSc, UMFPACK, and SuperLU only provide sequential implementations of the LU factorisation. If parallel

Table 10.3 Direct linear solve using different solver packages and ordering methods

Package	PETSc						SuperLU			
Ordering	None	ND	1WD	RCM	QMD	AMD	None	MMD A^TA	MMD A^T+A	COLAMD
MatGetOrdering	0.01	0.20	0.04	0.05	5.91	0.09	0.20	0.20	0.20	0.20
MatLUFactorSym	1.63	2.52	7.56	7.56	0.18	0.13	0	0	0	0
MatLUFactorNum	31.92	287.44	925.83	922.26	0.56	0.31	32.83	1.79	3.98	1.37
PCSetUp	33.56	290.16	933.43	929.87	6.65	0.53	33.03	1.99	4.18	1.57
PCApply	0.22	0.35	1.02	1.03	0.02	0.02	0.29	0.06	0.09	0.06
KSPSolve	33.78	290.51	934.45	930.90	6.67	0.55	33.32	2.05	4.27	1.63
nnz $(L+U)$	70.4M	110M	328M	328M	5.28M	4.67M	71.8M	8.21M	15.4M	8.46M
Fill ratio	35.03	54.89	163.10	163.26	2.63	2.32	35.72	4.08	7.68	4.21

Package	UMFPACK	MUMPS					SuperLU_Dist		
Ordering	AMD A^T+A	AMD	AMF	PORD	METIS	QAMD	None	MMD A^TA	MMD A^T+A
MatGetOrdering	0.20	0.20	0.20	0.20	0.20	0.20	0.20	0.20	0.20
MatLUFactorSym	0.48	0	0	0	0	0	0	0	0
MatLUFactorNum	0.90	1.26	1.21	2.07	1.90	1.27	84.07	2.23	1.27
PCSetUp	1.59	1.46	1.41	2.27	2.11	1.48	84.28	2.44	1.48
PCApply	0.03	0.12	0.12	0.12	0.12	0.13	0.66	0.17	0.13
KSPSolve	1.62	1.59	1.54	2.40	2.23	1.60	84.94	2.61	1.61
nnz $(L+U)$	4.63M	4.91M	4.64M	5.02M	5.56M	4.91M	70.8M	8.09M	5.26M
Fill ratio	2.30	2.44	2.31	2.50	2.77	2.44	35.24	4.02	2.61

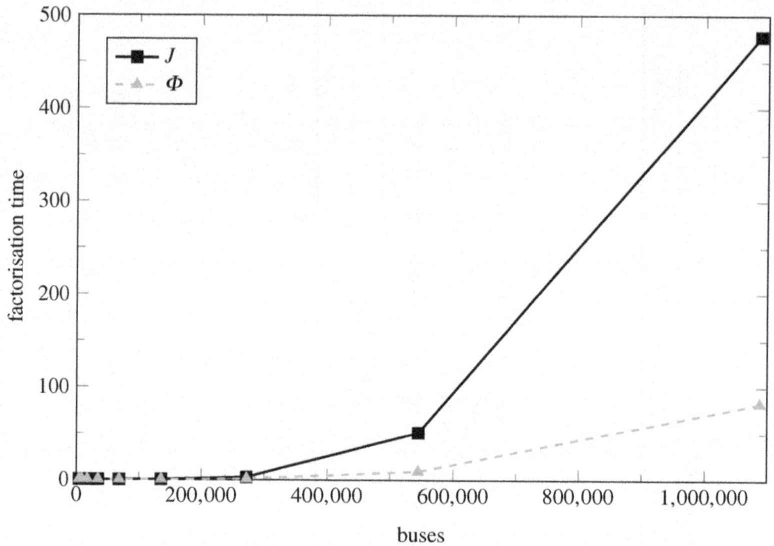

Fig. 10.1 LU factorisation of J and Φ

computing is desired, MUMPS and SuperLU_Dist can be used. For details on the tested ordering methods, see the manuals of the respective packages.

From the fill ratio, it is clear that AMD (and related methods) provide the best reordering in terms of fill-in. Using such a method reduces the fill ratio from a factor 35, to around 2.3. Note that some of the reordering methods lead to a fill ratio worse than that of the original ordering. These methods generally expect a much more structured matrix, as arises for example from the discretisation of differential equations on structured grids.

In terms of computational time, the PETSc solver with AMD ordering performs the best. Note that the difference with AMD ordering of other packages may well be due to the overhead of calling that package from PETSc. It also is possible to use external packages for the matrix reordering only, and then solve the problem with the PETSc solver. However, the results of using the AMD ordering within PETSc can be assumed to be representative of what modern packages have to offer.

Figure 10.1 shows the factorisation time for the first Jacobian matrix and the FDLF matrix Φ of the uctew032 test case, using the AMD reordering. The bad scaling of the LU decomposition is clearly visible. Recall from Sect. 8.2.1, that Φ consists of two independent blocks with each about a quarter of the nonzeros found in the Jacobian matrix. These blocks each have about half the dimension of the Jacobian matrix and can be factorised independently. As a result, the LU decomposition of Φ scales similar to that of the Jacobian matrix of a problem of half the size.

Table 10.4 Linear solve with ILU(8) and different orderings

Ordering	None	ND	1WD	RCM	QMD	AMD
MatGetOrdering	0.01	0.20	0.04	0.05	5.90	0.09
MatILUFactorSym	1.81	0.69	0.45	0.46	0.27	0.21
MatLUFactorNum	0.94	0.41	0.25	0.25	0.18	0.12
PCSetUp	2.76	1.31	0.74	0.77	6.36	0.42
PCApply	0.58	0.42	0.31	0.32	0.22	0.16
KSPSolve	3.54	2.01	1.29	1.32	6.76	0.74
GMRES iterations	12	15	13	13	11	10
nnz $(L+U)$	15.0M	7.62M	5.84M	5.96M	3.71M	3.56M
Fill ratio	7.47	3.79	2.90	2.97	1.85	1.77

10.1.2 ILU Factorisation

Table 10.4 shows results of numerical experiments regarding the effect of matrix reordering on the ILU(k) factorisation method. The first Jacobian system of the uctew032 test case is solved using GMRES, preconditioned with the ILU(8) factorisation of the coefficient matrix, using different ordering methods.

The fill ratio clearly illustrates that the ordering method influences the fill-in of the ILU(k) factorisation much less drastically than it did for the LU decomposition. Further note that all the tested reordering methods improve the fill-in compared to the natural ordering, whereas with the LU factorisation some orderings led to a higher fill ratio. However, the ordering methods that led to worse fill-in for the LU factorisation, here need more GMRES iterations to converge. This indicates that with these ordering methods, the ILU(8) preconditioner is of lesser quality than with the natural ordering.

The AMD ordering again performs the best of all available methods. It leads to the lowest fill ratio and the lowest amount of GMRES iterations at the same time. Thus it produces the highest quality preconditioner, with the lowest amount of nonzeros. Furthermore, both the calculation and the application of the AMD reordered preconditioner are faster than for any of the alternatives, thus leading to the fastest overall solution time.

Table 10.5 holds the results of numerical experiments with different ILU levels. Again, the first Jacobian system of the uctew032 test case is solved using ILU(k) preconditioned GMRES. The AMD reordering is used in all cases because it gives the best results, as was illustrated above for ILU(8) preconditioned GMRES. For comparison, the last column holds the equivalent data for a direct solve, which is also performed using AMD reordering.

The experiments suggest that using less than four levels leads to a preconditioner of too low quality. The factorisation is fast and the fill ratio is low, but due to the amount of GMRES iterations needed the solution time is still much higher than with more ILU levels. Using more than 16 levels leads to a very high quality preconditioner. With 64 levels the factorisation even becomes practically the same as the LU decomposition.

Table 10.5 Linear solve with different factorisations using AMD

Factorisation	ILU(0)	ILU(2)	ILU(4)	ILU(8)
Mat(I)LUFactorSym	0.08	0.14	0.17	0.21
MatLUFactorNum	0.08	0.10	0.11	0.12
PCSetUp	0.24	0.33	0.37	0.42
PCApply	2.30	0.71	0.35	0.16
KSPSolve	24.94	2.94	1.28	0.74
GMRES iterations	215	55	25	10
nnz $(L+U)$	2.01M	2.85M	3.21M	3.56M
Fill ratio	1	1.42	1.60	1.77

Factorisation	ILU(16)	ILU(32)	ILU(64)	LU
Mat(I)LUFactorSym	0.26	0.47	0.85	0.31
MatLUFactorNum	0.14	0.26	0.31	0.13
PCSetUp	0.50	0.81	1.25	0.53
PCApply	0.09	0.07	0.04	0.04
KSPSolve	0.67	0.94	1.31	0.59
GMRES iterations	5	3	1	1
nnz $(L+U)$	3.93M	4.52M	4.67M	4.67M
Fill ratio	1.96	2.25	2.32	2.32

Few GMRES iterations are needed to solve the linear problem, but the factorisation takes more time and the fill ratio is larger, making the overall solution time higher than with 4–16 levels.

Figure 10.2 shows the scaling behaviour of different ILU(k) levels. The ILU(2) and ILU(8) factorisations both scale very well in the problem size, but solving the linear problem with ILU(8) is approximately twice as fast as with ILU(2). Higher ILU levels start to lose the linear scaling behaviour, as illustrated by the ILU(32) graph, which is nearing that of the LU factorisation.

Figure 10.3 inspects the scaling of ILU factorisations with around eight levels. ILU(4), ILU(8), and ILU(12) all scale approximately linearly, with ILU(4) being slightly slower than the other two. The graphs of ILU(8) and ILU(12) are practically on top of each other. ILU(16) is still very fast, but no longer scales linearly. Therefore, in the remainder of this chapter only 4, 8, and 12 levels of ILU factorisations are considered.

10.2 Forcing Terms

In this section, the behaviour is explored of the Dembo and Steihaug [9] and the Eisenstat and Walker [10] forcing term strategies, that were presented in Sect. 8.3. Their performance is illustrated by analysing a representative test case, solved with

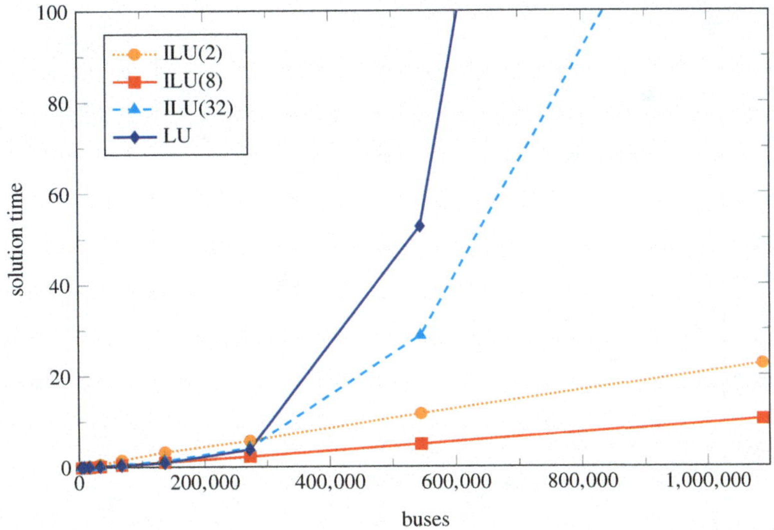

Fig. 10.2 Linear solve for different problem sizes

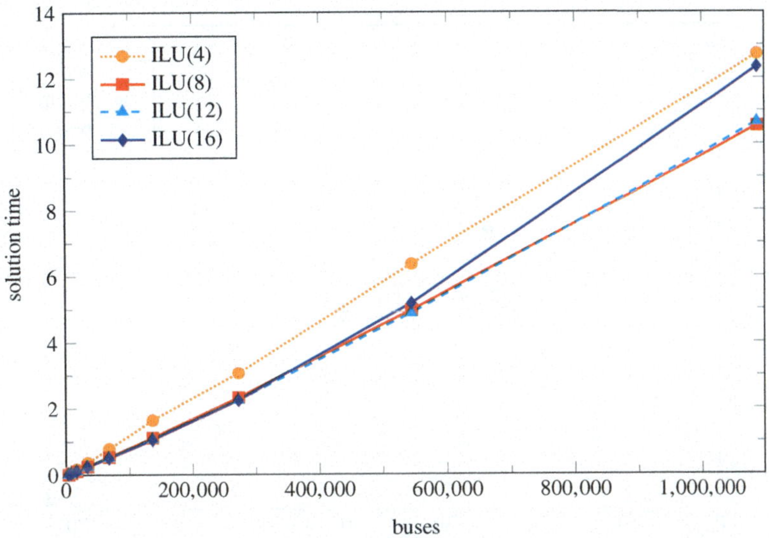

Fig. 10.3 Linear solve for different problem sizes

both forcing term strategies. To this end, the uctew032 test problem is solved using GMRES, preconditioned with the ILU(8) factorisation of the FDLF matrix Φ.

Figures 10.4 and 10.5 show the resulting nonlinear convergence in the total number of GMRES iterations, for Dembo and Steihaug and Eisenstat and Walker forcing terms respectively. The legend of these figures is explained in Table 10.6.

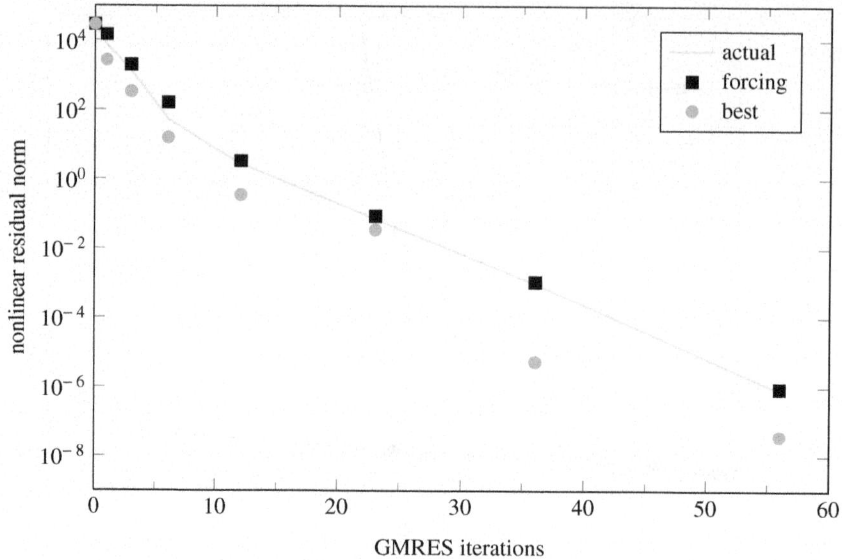

Fig. 10.4 Dembo and Steihaug forcing terms

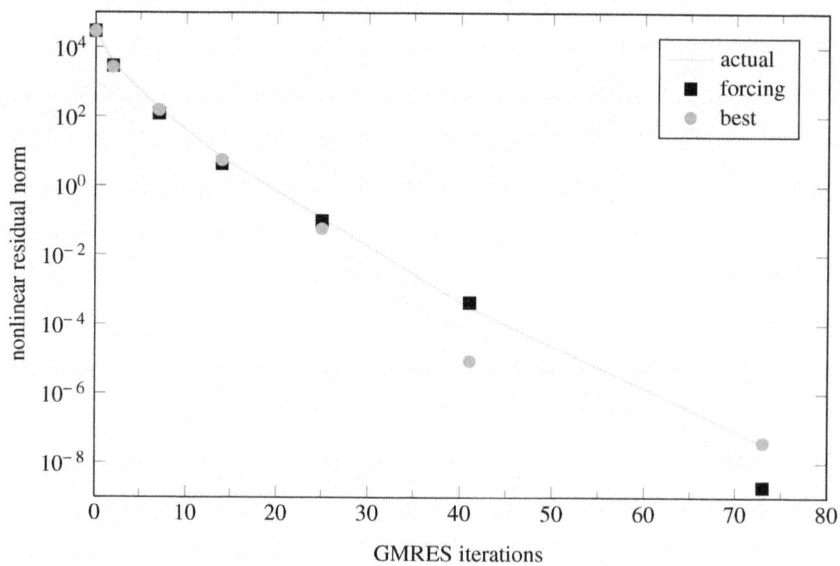

Fig. 10.5 Eisenstat and Walker forcing terms

If the forcing residual norm is below the actual residual norm, then the method is oversolving in that iteration. If the best residual norm is below the actual residual norm, then the method is undersolving. Note that when there is no oversolving, the

Table 10.6 Nonlinear residual norm legend

actual	: Actual nonlinear residual norm in each Newton iteration
forcing	: Nonlinear residual norm resulting from multiplying the previous nonlinear residual norm by the forcing term used in the current Newton iteration
best	: Nonlinear residual norm resulting from doing a full accuracy linear solve in the current Newton iteration

forcing residual norm is mostly on top of the actual residual norm, as expected from the theory in Chap. 5.

Using this knowledge, the following conclusions can be drawn for the presented experiments. The Dembo and Steihaug forcing terms lead to undersolving. The Eisenstat and Walker forcing terms are spot-on for the first 4 Newton iterations, while there is some undersolving in iteration 5, and oversolving in iteration 6. These conclusions are consistent with the general behaviour of these two forcing terms strategies in our power flow experiments, see Sect. 10.3.

The Eisenstat and Walker forcing terms include in their calculation the nonlinear residual norm of the current and previous Newton iterations, and the latest linear residual norm, all the ingredients needed to determine whether there was undersolving or oversolving in the previous iteration. Using this information, the method tends to correct itself. If there is a lot of oversolving in the previous iteration, then the forcing term will be chosen larger, and vice versa. Therefore, the Eisenstat and Walker strategy can adapt to the problem very well.

10.3 Power Flow

This section reports on numerical experiments with the Newton-Krylov power flow solver, solving the full power flow problem. In each Newton iteration, the linear system is solved using preconditioned GMRES. LU, ILU(4), ILU(8), and ILU(12) factorisations of the target matrices J_i, J_0, and Φ are all tested as preconditioner. The problems are solved from a flat start, up to an accuracy of 10^{-6} p.u. No solution adjustments, like reactive power limits or tap changing, are used in the experiments.

Tables 10.7 and 10.8 show the results for Dembo and Steihaug forcing terms and Eisenstat and Walker forcing terms respectively. For each experiment, the number of Newton iterations p and GMRES iterations q are presented in the format p/q, and the solution time is given in seconds. In both tables, the Newton iteration count and solution time when using a direct solver are added as a reference.

For the smaller test cases, the best results are generally obtained when using an LU decomposition of J_0 as preconditioner. For the largest problems, the LU decomposition becomes too slow due to the bad scaling, as demonstrated in Sect. 10.1.1. For these cases, ILU(12) factorisations of J_0 and Φ give the best results.

Table 10.7 Power flow experiments using the Dembo and Steihaug forcing terms

Power flow problem			uctew001	uctew002	uctew004	uctew008	uctew016	uctew032	uctew064	uctew128	uctew256
Direct		Iterations	7	7	7	7	7	7	7	6	8
		Time	0.077	0.16	0.33	0.69	1.54	3.96	20.89	305.41	3810.80
J_i	ILU(4)	Iterations	7/76	7/79	7/75	7/79	7/80	7/77	7/72	8/108	8/107
		Time	0.12	0.26	0.53	1.14	2.40	4.81	9.43	27.09	54.91
J_i	ILU(8)	Iterations	7/43	7/47	7/38	7/42	7/39	7/39	7/39	8/46	9/49
		Time	0.10	0.21	0.42	0.89	1.76	3.62	7.42	18.43	42.78
J_i	ILU(12)	Iterations	7/24	7/28	7/27	7/27	7/26	7/24	7/24	7/22	9/32
		Time	0.089	0.19	0.39	0.81	1.64	3.27	6.75	15.07	42.25
J_0	LU	Iterations	7/23	7/22	7/22	7/19	7/21	7/23	7/28	8/40	9/47
		Time	0.068	0.14	0.28	0.57	1.21	2.70	7.89	65.13	517.90
J_0	ILU(4)	Iterations	7/83	7/84	7/81	7/79	7/82	8/136	8/118	8/128	8/135
		Time	0.11	0.23	0.49	1.01	2.14	7.08	12.56	27.73	58.90
J_0	ILU(8)	Iterations	7/40	7/38	7/35	7/35	8/58	8/55	8/68	8/72	9/88
		Time	0.082	0.16	0.33	0.68	1.84	3.66	8.51	18.03	42.93
J_0	ILU(12)	Iterations	7/27	7/27	7/24	7/24	7/24	8/36	7/32	8/56	9/65
		Time	0.073	0.15	0.30	0.62	1.25	3.08	5.75	16.09	36.87
Φ	LU	Iterations	7/34	7/33	7/33	7/33	7/35	7/33	7/35	8/59	8/56
		Time	0.081	0.13	0.28	0.58	1.22	2.54	5.84	23.22	113.18
Φ	ILU(4)	Iterations	7/107	7/101	8/160	8/161	8/164	8/162	8/142	8/146	8/178
		Time	0.12	0.23	0.78	1.68	3.68	7.77	13.76	29.29	73.77
Φ	ILU(8)	Iterations	7/61	7/54	7/50	7/52	7/56	7/55	8/93	8/95	8/94
		Time	0.083	0.16	0.32	0.70	1.49	3.11	9.65	20.08	40.20
Φ	ILU(12)	Iterations	7/44	7/40	7/40	7/41	7/44	7/45	7/47	8/60	8/69
		Time	0.073	0.14	0.30	0.63	1.32	2.81	5.97	14.57	32.57

Table 10.8 Power flow experiments using the Eisenstat and Walker forcing terms

Power flow problem			uctew001	uctew002	uctew004	uctew008	uctew016	uctew032	uctew064	uctew128	uctew256
Direct		Iterations	7	7	7	7	7	7	7	6	8
		Time	0.077	0.16	0.33	0.69	1.54	3.96	20.89	305.41	3810.80
J_i	ILU(4)	Iterations	7/102	7/113	7/116	7/123	7/128	7/127	7/117	7/115	7/114
		Time	0.15	0.34	0.74	1.65	3.56	7.34	13.87	28.10	57.74
J_i	ILU(8)	Iterations	7/44	7/44	7/53	7/50	7/50	7/50	7/48	7/40	9/50
		Time	0.10	0.21	0.48	0.96	1.97	4.04	8.15	16.55	43.42
J_i	ILU(12)	Iterations	7/25	7/28	7/27	7/31	7/32	7/32	7/33	7/29	8/22
		Time	0.090	0.19	0.39	0.84	1.75	3.56	7.45	16.18	36.57
J_0	LU	Iterations	6/18	6/18	6/18	6/18	7/29	6/21	6/26	7/40	8/37
		Time	**0.060**	**0.12**	**0.25**	**0.52**	1.34	**2.50**	7.51	64.55	511.08
J_0	ILU(4)	Iterations	7/112	7/121	7/126	7/128	7/141	7/167	7/139	7/150	7/165
		Time	0.14	0.33	0.72	1.54	3.55	8.95	15.30	33.16	76.14
J_0	ILU(8)	Iterations	7/53	7/59	7/55	7/56	7/61	7/63	7/73	7/73	8/67
		Time	0.092	0.20	0.41	0.86	1.83	3.88	8.73	18.16	35.78
J_0	ILU(12)	Iterations	7/39	7/43	6/26	6/26	6/28	6/30	6/35	7/52	8/50
		Time	0.082	0.18	0.29	0.60	1.24	2.61	**5.69**	15.01	31.61
Φ	LU	Iterations	6/35	6/35	6/35	6/36	6/37	6/40	6/44	6/49	6/54
		Time	0.063	0.13	0.26	0.56	1.18	2.62	6.16	20.39	110.05
Φ	ILU(4)	Iterations	6/119	6/115	6/121	6/137	6/143	6/141	6/135	6/145	6/152
		Time	0.12	0.24	0.57	1.40	3.17	6.76	13.36	29.42	62.61
Φ	ILU(8)	Iterations	6/68	6/71	6/64	6/70	6/70	6/75	6/89	6/90	6/91
		Time	0.085	0.18	0.35	0.79	1.66	3.67	9.07	18.79	37.58
Φ	ILU(12)	Iterations	6/55	6/55	6/54	6/60	5/41	6/58	6/68	6/62	6/61
		Time	0.076	0.15	0.32	0.72	**1.15**	3.07	7.16	**13.69**	**27.54**

The Eisenstat and Walker forcing terms perform slightly better than the Dembo and Steihaug strategy. With Dembo and Steihaug forcing terms a higher amount of Newton iterations is needed in many cases. This suggests undersolving, which was also observed for this strategy in Sect. 10.2.

For each individual test case, the smallest solution time is obtained using some J_0 or Φ based preconditioner together with the Eisenstat and Walker forcing terms. The best solution time for each test case is marked in the table with a bold numbers. Note that in all the cases where a preconditioner based on Φ gave the best result, this is due to the J_0 based alternative needing one or two extra Newton iterations. As the forcing terms are not directly dependent on the quality of the preconditioner, these extra Newton iterations are mostly due to being a bit lucky, or unlucky, with how the inexact Newton steps turn out exactly.

10.3.1 Scaling

The number of Newton iterations needed to solve the problem generally does not increase much when the test case size increases, see again Tables 10.7 and 10.8. For some combinations of preconditioner and forcing terms the largest problems require a few more iterations, but for other combinations the number of Newton iterations stays the same. This suggests that an increased number of iterations is more due to getting a bit unlucky with the Newton steps, than it being a direct consequence of the increased problem size.

Similarly, for each combination of preconditioner and forcing terms, the total number of GMRES iterations is fairly constant for increasing problem sizes. When a significantly higher amount of GMRES iterations is needed, it is generally due to an extra Newton iteration being used.

The basic operations of the Newton-Krylov power flow solver, are all vector operations and operations on the Jacobian matrix. A larger power system has more buses, but generally not more connections per bus. As a result, the number of nonzeros per row in the Jacobian matrix does not grow for larger problems. Thus the total number of nonzeros scales linearly in the problem size, and so does the computation time of the needed operations on the Jacobian matrix, like matvecs. The exception is the factorisation operation, see Sect. 10.1.

When the number of Newton and GMRES iterations is constant in the problem size, the scaling of the Newton-Krylov power flow solver can therefore be expected to be linear, as long as the factorisation also scales linearly. If the factorisation scales badly, as with the LU decomposition, the power flow solver also scales badly.

Figures 10.6, 10.7, and 10.8 show the scaling of the computational time when using different factorisations of J_i, J_0, and Φ as preconditioner, respectively. Indeed, the solver exhibits approximately linear scaling when using ILU(k) factorisations with 4–12 levels, which were shown to scale approximately linearly up to a million buses in Sect. 10.1.2. Using the LU factorisation leads to bad scaling, as expected from the results of Sect. 10.1.1.

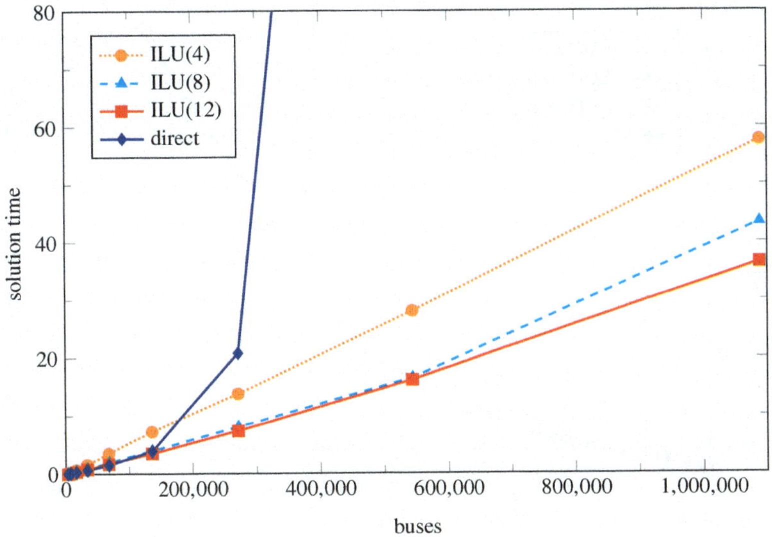

Fig. 10.6 Power flow with J_i based preconditioning

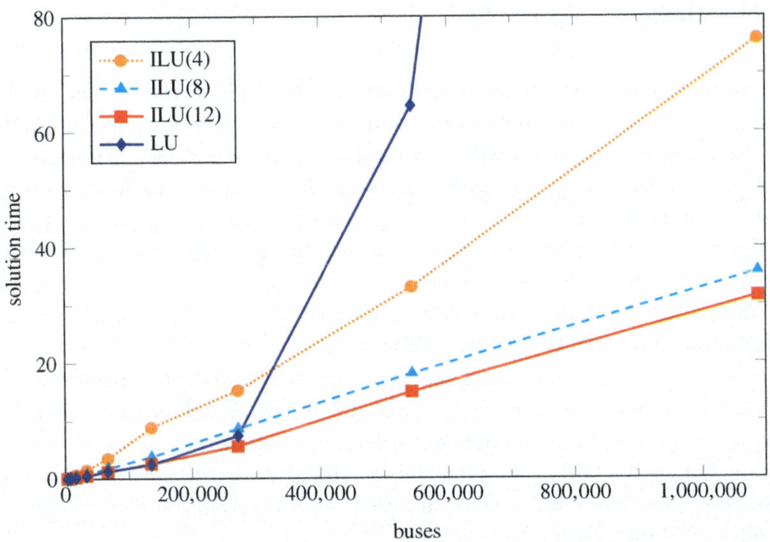

Fig. 10.7 Power flow with J_0 based preconditioning

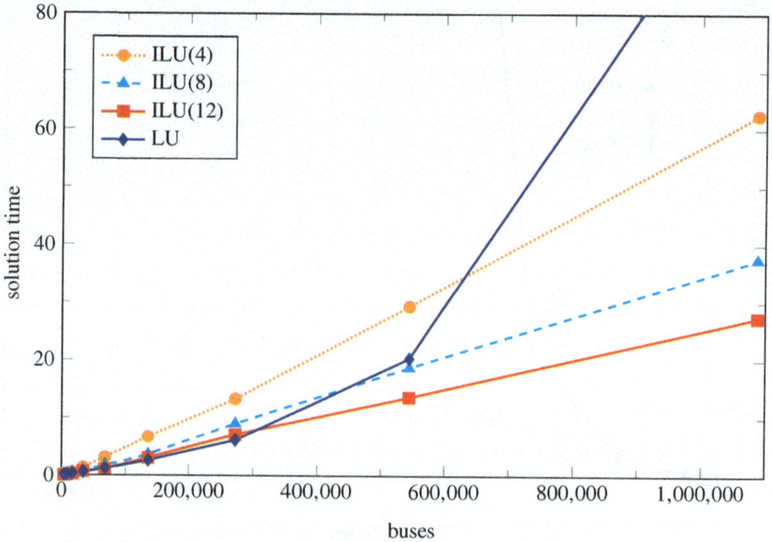

Fig. 10.8 Power flow with Φ based preconditioning

10.4 Contingency Analysis

This section treats numerical experiments with the Newton-Krylov power flow solver, applied to the contingency analysis problem (see Sect. 6.4 and Chap. 9). The UCTE winter 2008 study model (uctew001) is used as the base test case. The contingency cases consist of the base case, with a single pair of buses (which were connected in the base case) disconnected, simulating branch outages. This makes 6784 contingency cases, of which 95 cases are not solved because they contain disconnected subnetworks, leaving 6689 contingency cases. The base case power flow problem is solved first, after which the power flow problem of each of the contingency cases is solved, making a total of 6690 power flow solves.

Table 10.9 presents results using Eisenstat and Walker forcing terms. A maximum of 12 Newton iterations is allowed per case, and no line search is used to keep results as transparent as possible. All cases are solved up to an accuracy of 10^{-4} p.u.

The top half of the table uses a flat start for all cases, while the bottom half solves only the base case using a flat start, and then uses the solution of the base case as initial iterate for the contingency cases.

The left column shows the results using classical Newton power flow with a direct linear solver.

The middle column solves each case with Newton-GMRES, preconditioned with the LU factorisation of the initial Jacobian of that case, which was shown to be the fastest option for the base case in Sect. 10.3.

The right column again uses the Newton-Krylov power flow solver, but preconditions GMRES with the LU factorisation of the base case Jacobian evaluated in

Table 10.9 Contingency analysis using Eisenstat and Walker forcing terms

Initial solution	Flat start					
Preconditioning	Direct		Own J_0		Base J_0	
	Count	Iter	Count	Iter	Count	Iter
Converged	6665	7/7	6665	6/15	6666	6/20
Diverged	24	12/12	24	12/73	23	12/88
	Count	Time	Count	Time	Count	Time
PCSetUp	46948	191	6690	57.6	2	0.02
PCApply	46948	16.2	142263	48.9	176899	62.0
KSPSolve	46948	208	40287	135	40360	99.8
CalcJac	53638	98.9	46977	86.2	47050	86.2
CA	1	320	1	238	1	198

Initial solution	Base case solution					
Preconditioning	Direct		Own J_0		Base J^*	
	Count	Iter	Count	Iter	Count	Iter
Converged	6666	2.2/2.2	6666	2.3/3.3	6665	2.4/6.3
Diverged	23	12/12	23	12/73	24	12/88
	Count	Time	Count	Time	Count	Time
PCSetUp	14975	85.0	6686	57.8	2	0.02
PCApply	14975	5.18	38335	13.2	60661	21.3
KSPSolve	14975	90.3	15472	77.6	16418	33.5
CalcJac	21665	43.0	22162	42.1	23108	43.7
CA	1	140	1	132	1	84.4

the initial iterate used for the contingency cases. With a flat start, the contingency cases are thus preconditioned with the initial Jacobian J_0 of the base case. And when starting with the solution of the base case, they are preconditioned with the base case Jacobian evaluated in the base case solution, denoted by J^*.

The converged and diverged rows, show the number of contingency cases that converge and diverge respectively, and the average amount of Newton iterations and GMRES iterations per case. Diverging contingency cases is a common problem in contingency analysis, that we will not go into further here. Note that one case is solved with some methods, but fails to converge with others.

For an explanation of PCSetUp, PCApply, and KSPSolve see Table 10.2 (p. 88). CalcJac stands for the calculation of the Jacobian system, i.e., the power mismatch vector and the Jacobian matrix. The abbreviation CA is used for the computational time of the entire contingency analysis process.

No results are presented for the Dembo and Steihaug forcing term strategy, as the Eisenstat and Walker forcing terms perform significantly better, especially when using the base case solution as initial solution for the contingency cases. The adaptive nature of the Eisenstat and Walker forcing terms makes them very well-suited to handle the resulting varying initial residual norms.

Two ideas to improve on these forcing terms are worth mentioning. One is to reduce the initial forcing term value of the Eisenstat and Walker strategy when using the base case solution as initial iterate. Because this initial iterate is generally much closer to the solution than a flat start, it is expected that a greater improvement can be attained in the first Newton iteration than the default $\eta_0 = 0.1$ that is used for the Eisenstat and Walker strategy. The other is to log the convergence of the base case, and use this as a model for the expected convergence of the contingency cases. For our test cases, both methods provide only very minor improvements over using plain Eisenstat and Walker forcing terms.

References

1. Balay, S., Buschelman, K., Eijkhout, V., Gropp, W.D., Kaushik, D., Knepley, M.G., Curfman McInnes, L., Smith, B.F., Zhang, H.: PETSc users manual. Tech. Rep. ANL-95/11 - Revision 3.1, Argonne National Laboratory (2010). http://www.mcs.anl.gov/petsc/
2. Duff, I.S., Erisman, A.M., Reid, J.K.: Direct Methods for Sparse Matrices. Oxford University Press, New York (1986)
3. Davis, T.A.: Direct Methods for Sparse Linear Systems. SIAM, Philadelphia (2006)
4. Amestoy, P.R., Davis, T.A., Duff, I.S.: An approximate minimum degree ordering algorithm. SIAM J. Matrix Anal. Appl. **17**(4), 886–905 (1996)
5. Davis, T.A.: A column pre-ordering strategy for the unsymmetric-pattern multifrontal method. ACM Trans. Math. Softw. **30**(2), 165–195 (2004)
6. Amestoy, P.R., Duff, I.S., L'Excellent, J.Y., Koster, J.: A fully asynchronous multifrontal solver using distributed dynamic scheduling. SIAM J. Matrix Anal. Appl. **23**(1), 15–41 (2001)
7. Demmel, J.W., Eisenstat, S.C., Gilbert, J.R., Li, X.S., Liu, J.W.H.: A supernodal approach to sparse partial pivoting. SIAM J. Matrix Anal. Appl. **20**(3), 720–755 (1999)
8. Li, X.S., Demmel, J.W.: SuperLU_DIST: a scalable distributed-memory sparse direct solver for unsymmetric linear systems. ACM Trans. Math. Softw. **29**(2), 110–140 (2003)
9. Dembo, R.S., Steihaug, T.: Truncated-Newton algorithms for large-scale unconstrained optimization. Math. Prog. **26**, 190–212 (1983)
10. Eisenstat, S.C., Walker, H.F.: Choosing the forcing terms in an inexact Newton method. SIAM J. Sci. Comput. **17**(1), 16–32 (1996)

Chapter 11
Power Flow Test Cases

To conduct numerical experiments with power flow solvers, a test set of power flow problems is needed. Problems with up to a few hundred buses are readily available, but problems of realistic size are hard to come by. Transmission systems are vital to society, and can be vulnerable to attacks if the attackers know where to strike. Therefore, models of the actual systems are not publicly available.

For our research, we were able to use the UCTE[1] winter 2008 study model, which consists of 4253 buses and 7191 lines. Larger test cases were constructed by copying the model and interconnecting the copies with new transmission lines, as detailed in Sect. 11.1. This proved to give better results than generating realistic models of virtual power systems from scratch.

The test cases are named uctewXXX, where XXX is the number of copies of the original problem used in the construction of the test case. Table 11.1 shows the number of buses, branches, and nonzeros in the Jacobian matrix, for each test case. These test cases were also used in [1, 2].

11.1 Construction

The uctew001 test case is simply the UCTE winter 2008 study model. Each next test case is constructed by connecting two copies of the previous test case. The important choices in this process are: which of the buses to connect, how to connect them, and how to deal with the slack buses. Figure 11.1 shows a schematic representation of the construction process used for our test cases.

The two network copies A and B each have their own slack bus, denoted by As and Bs respectively. If one slack bus is simply removed, together with all branches

[1] UCTE is a former association of transmission system operators in Europe. As of July 2009, the European Network of Transmission System Operators for Electricity (ENTSO-E), a newly formed association of 41 TSOs from 34 countries in Europe, has taken over all operational tasks of the existing European TSO associations, including UCTE. See http://www.entsoe.eu/.

R. Idema and D. J. P. Lahaye, *Computational Methods in Power System Analysis*, Atlantis Studies in Scientific Computing in Electromagnetics,
DOI: 10.2991/978-94-6239-064-5_11,

Table 11.1 Power flow test cases for numerical experiments

Name	Buses	Branches	nnz (J)
uctew001	4,253	7,191	62,654
uctew002	8,505	14,390	125,372
uctew004	17,009	28,796	250,872
uctew008	34,017	57,624	502,000
uctew016	68,033	115,312	1,004,512
uctew032	136,065	230,752	2,010,048
uctew064	272,129	461,760	4,022,144
uctew128	544,257	924,032	8,048,384
uctew256	1,088,513	1,849,088	16,104,960

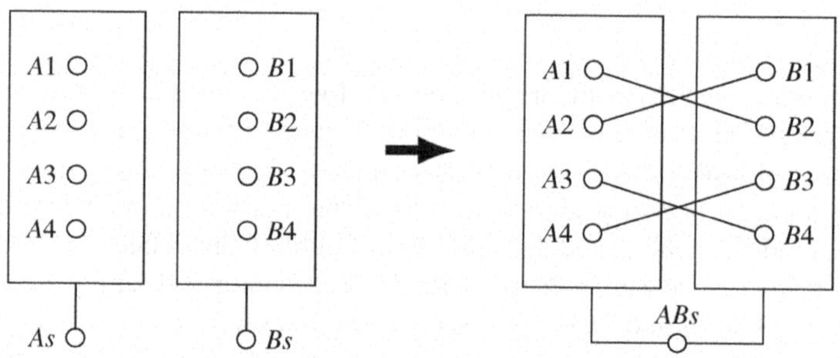

Fig. 11.1 Test case construction process

connected to it, all the generation in that slack bus has to be provided for by the other slack bus. Because the other slack bus is in a totally different area of the network, this may lead to an imbalanced test case. Therefore, it is better to combine the two slack buses into one new slack bus *ABs* that is connected to all the buses that either of the old slack buses was connected to.

When existing power systems are connected in practice, the network connection is generally made at the highest voltage level. Thus it makes sense to do the same when constructing test cases by connecting existing networks. For our test cases, a number of load buses at the highest voltage level is selected, approximately uniformly distributed by bus index, with a small random element.

Connecting completely different regions of the network copies may lead to a serious imbalance. Thus, each bus in A should be connected to a bus in B that corresponds to a nearby bus in A. If each bus is connected directly to its corresponding bus in the other network, no current will flow between A and B. The solution of the newly constructed problem would simply consists of the original network solution in both A and B. Therefore, in our test cases the buses are connected per pair $A1$ and $A2$, close to each other, to the corresponding buses $B1$ and $B2$ in the other network, such that $A1$ is connected to $B2$, and $A2$ is connected to $B1$.

The number of connecting branches between the two network copies is also of importance. In our test cases the number of buses selected in A is 8 times the amount of original networks incorporated in A. If too few cross branches are added, the networks A and B are nearly decoupled. This results in an admittance matrix with two blocks on nonzeros on the diagonal, and only a few nonzeros outside of these blocks. This structure continues into the Jacobian matrix, and factorising such a Jacobian is similar to factorising the two diagonal blocks independently. Any issues with the scaling of the factorisation method would be lost for such a test case.

References

1. Idema, R., Lahaye, D.J.P., Vuik, C., van der Sluis, L.: Scalable Newton-Krylov solver for very large power flow problems. IEEE Trans. Power Syst. **27**(1), 390–396 (2012)
2. Idema, R., Papaefthymiou, G., Lahaye, D.J.P., Vuik, C., van der Sluis, L.: Towards faster solution of large power flow problems. IEEE Trans. Power Syst. **28**(4), 4918–4925 (2013)

Index

R. Idema and D. J. P. Lahaye, *Computational Methods in Power System Analysis*,
Atlantis Studies in Scientific Computing in Electromagnetics,
DOI: 10.2991/978-94-6239-064-5,

The manufacturer's authorised representative in the EU is Springer Nature Customer Service Centre GmbH, Europaplatz 3, 69115 Heidelberg, Germany. If you have any concerns regarding our products, please contact ProductSafety@springernature.com

Printed and bound by CPI Group (UK) Ltd, Croydon, CR0 4YY
22/09/2026
02221057-0017